Steve Donnell

An Introduction to
LINEAR ALGEBRA

15-01

pub. desc.

An Introduction to

LINEAR ALGEBRA

MR 3676053

Ravi P. Agarwal and Cristina Flaut

CRC Press
Taylor & Francis Group
Boca Raton London New York

CRC Press is an imprint of the
Taylor & Francis Group, an **Informa** business

A CHAPMAN & HALL BOOK

CRC Press
Taylor & Francis Group
6000 Broken Sound Parkway NW, Suite 300
Boca Raton, FL 33487-2742

© 2017 by Taylor & Francis Group, LLC
CRC Press is an imprint of Taylor & Francis Group, an Informa business

No claim to original U.S. Government works

Printed on acid-free paper

International Standard Book Number-13: 978-1-138-62670-6 (Hardback)

Visit the Taylor & Francis Web site at
http://www.taylorandfrancis.com

and the CRC Press Web site at
http://www.crcpress.com

Dedicated to our mothers:
Godawari Agarwal, Elena Paiu, and Maria Paiu

Contents

Preface

Linear algebra is a branch of both pure and applied mathematics. It provides the foundation for multi-dimensional representations of mathematical reasoning. It deals with systems of linear equations, matrices, determinants, vectors and vector spaces, transformations, and eigenvalues and eigenvectors. The techniques of linear algebra are extensively used in every science where often it becomes necessary to approximate nonlinear equations by linear equations. Linear algebra also helps to find solutions for linear systems of differential and difference equations. In pure mathematics, linear algebra (particularly, vector spaces) is used in many different areas of algebra such as group theory, module theory, representation theory, ring theory, Galöis theory, and this list continues. This has given linear algebra a unique place in mathematics curricula all over the world, and it is now being taught as a compulsory course at various levels in almost every institution.

Although several fabulous books on linear algebra have been written, the present rigorous and transparent introductory text can be used directly in class for students of applied sciences. In fact, in an effort to bring the subject to a wider audience, we provide a compact, but thorough, introduction to the subject in **An Introduction to Linear Algebra**. This book is intended for senior undergraduate and for beginning graduate one-semester courses.

The subject matter has been organized in the form of theorems and their proofs, and the presentation is rather unconventional. It comprises 25 class-tested lectures that the first author has given to math majors and engineering students at various institutions over a period of almost 40 years. It is our belief that the content in a particular chapter, together with the problems therein, provides fairly adequate coverage of the topic under study.

A brief description of the topics covered in this book follows: In **Chapter 1**, we define axiomatically terms such as field, vector, vector space, subspace, linear combination of vectors, and span of vectors. In **Chapter 2**, we introduce various types of matrices and formalize the basic operations: matrix addition, subtraction, scalar multiplication, and matrix multiplication. We show that the set of all $m \times n$ matrices under the operations matrix addition and scalar multiplication is a vector space. In **Chapter 3**, we begin with the definition of a determinant and then briefly sketch the important properties of determinants. In **Chapter 4**, we provide necessary and sufficient conditions for a square matrix to be invertible. We shall show that the theory of determinants can be applied to find an analytical representation of the inverse of a square matrix. Here we also use elementary theory of difference equations to find inverses of some band matrices.

The main purpose of **Chapters 5 and 6** is to discuss systematically Gauss and Gauss–Jordan elimination methods to solve m linear equations in n unknowns. These equations are conveniently written as $Ax = b$, where A is an $m \times n$ matrix, x is an $n \times 1$ unknown vector, and b is an $m \times 1$ vector. For this, we introduce the terms consistent, inconsistent, solution space, null space, augmented matrix, echelon form of a matrix, pivot, elementary row operations, elementary matrix, row equivalent matrix, row canonical form, and rank of a matrix. These methods also provide effective algorithms to compute determinants and inverses of matrices. We also prove several theoretical results that yield necessary and sufficient conditions for a linear system of equations to have a solution. **Chapter 7** deals with a modified but restricted realization of Gaussian elimination. We factorize a given $m \times n$ matrix A to a product of two matrices L and U, where L is an $m \times m$ lower triangular matrix, and U is an $m \times n$ upper triangular matrix. Here we also discuss various variants and applications of this factorization.

In **Chapter 8**, we define the concepts linear dependence and linear independence of vectors. These concepts play an essential role in linear algebra and as a whole in mathematics. Linear dependence and independence distinguish between two vectors being essentially the same or different. In **Chapter 9**, for a given vector space, first we introduce the concept of a basis and then describe its dimension in terms of the number of vectors in the basis. Here we also introduce the concept of direct sum of two subspaces. In **Chapter 10**, we extend the known geometric interpretation of the coordinates of a vector in R^3 to a general vector space. We show how the coordinates of a vector space with respect to one basis can be changed to another basis. Here we also define the terms ordered basis, isomorphism, and transition matrix. In **Chapter 11**, we redefine rank of a matrix and show how this number is directly related to the dimension of the solution space of homogeneous linear systems. Here for a given matrix we also define row space, column space, left and right inverses, and provide necessary and sufficient conditions for their existence. In **Chapter 12**, we introduce the concept of linear mappings between two vector spaces and extend some results of earlier chapters. In **Chapter 13**, we establish a connection between linear mappings and matrices. We also introduce the concept of similar matrices, which plays an important role in later chapters. In **Chapter 14**, we extend the familiar concept inner product of two or three dimensional vectors to general vector spaces. Our definition of inner products leads to the generalization of the notion of perpendicular vectors, called orthogonal vectors. We also discuss the concepts projection of a vector onto another vector, unitary space, orthogonal complement, orthogonal basis, and Fourier expansion. This chapter concludes with the well-known Gram–Schmidt orthogonalization process. In **Chapter 15**, we discuss a special type of linear mapping, known as linear functional. We also address such notions as dual space, dual basis, second dual, natural mapping, adjoint mapping, annihilator, and prove the famous Riesz representation theorem.

Chapter 16 deals with the eigenvalues and eigenvectors of matrices. We summarize those properties of the eigenvalues and eigenvectors of matrices that facilitate their computation. Here we come across the concepts characteristic polynomial, algebraic and geometric multiplicities of eigenvalues, eigenspace, and companion and circulant matrices. We begin **Chapter 17** with the definition of a norm of a vector and then extend it to a matrix. Next, we drive some estimates on the eigenvalues of a given matrix, and prove some useful convergence results. Here we also establish well known Cauchy–Schwarz, Minkowski, and Bessel inequalities, and discuss the terms spectral radius, Rayleigh quotient, and best approximation.

In **Chapter 18**, we show that if algebraic and geometric multiplicities of an $n \times n$ matrix A are the same, then it can be diagonalized, i.e., $A = PDP^{-1}$; here, P is a nonsingular matrix and D is a diagonal matrix. Next, we provide necessary and sufficient conditions for A to be orthogonally diagonalizable, i.e., $A = QDQ^t$, where Q is an orthogonal matrix. Then, we discuss QR factorization of the matrix A. We also furnish complete computationable characterizations of the matrices P, D, Q, and R. In **Chapter 19**, we develop a generalization of the diagonalization procedure discussed in Chapter 18. This factorization is applicable to any real $m \times n$ matrix A, and in the literature has been named singular value decomposition. Here we also discuss reduced singular value decomposition.

In **Chapter 20**, we show how linear algebra (especially eigenvalues and eigenvectors) plays an important role to find the solutions of homogeneous differential and difference systems with constant coefficients. Here we also develop continuous and discrete versions of the famous Putzer's algorithm. In a wide range of applications, we encounter problems in which a given system $Ax = b$ does not have a solution. For such a system we seek a vector(s) $\hat{x}$ so that the error in the Euclidean norm, i.e., $\|A\hat{x} - b\|_2$, is as small as possible (minimized). This solution(s) $\hat{x}$ is called the least squares approximate solution. In **Chapter 21**, we shall show that a least squares approximate solution always exists and can be conveniently computed by solving a related system of n equations in n unknowns (normal equations). In **Chapter 22**, we study quadratic and diagonal quadratic forms in n variables, and provide criteria for them to be positive definite. Here we also discuss maximum and minimum of the quadratic forms subject to some constraints (constrained optimization). In **Chapter 23**, first we define positive definite symmetric matrices in terms of quadratic forms, and then for a symmetric matrix to be positive definite, we provide necessary and sufficient conditions. Next, for a symmetric matrix we revisit LU-factorization, and give conditions for a unique factorization LDL^t, where L is a lower triangular matrix with all diagonal elements 1, and D is a diagonal matrix with all positive elements. We also discuss Cholesky's decomposition $L_c L_c^t$ where $L_c = LD^{1/2}$, and for its computation provide Cholesky's algorithm. This is followed by Sylvester's criterion, which gives easily verifiable necessary and sufficient conditions for a symmetric matrix to be positive definite. We conclude this chapter with a polar decomposition. In **Chapter 24**, we introduce the concept of pseudo/generalized (Moore–Penrose) inverse which is applicable to all $m \times n$ matrices. As an illustration we apply Moore–Penrose inverse to least squares solutions of linear equations. Finally, in **Chapter 25**, we briefly discuss irreducible, nonnegative, diagonally dominant, monotone, and Toeplitz matrices. We state 11 theorems which, from the practical point of view, are of immense value. These types of matrices arise in several diverse fields, and hence have attracted considerable attention in recent years.

In this book, there are 148 examples that explain each concept and demonstrate the importance of every result. Two types of 254 problems are also included, those that illustrate the general theory and others designed to fill out text material. The problems form an integral part of the book, and every reader is urged to attempt most, if not all of them. For the convenience of the reader, we have provided answers or hints to all the problems.

In writing a book of this nature, no originality can be claimed, only a humble attempt has been made to present the subject as simply, clearly, and accurately as possible. The illustrative examples are usually very simple, keeping in mind an average student.

It is earnestly hoped that **An Introduction to Linear Algebra** will serve an inquisitive reader as a starting point in this rich, vast, and ever-expanding field of knowledge.

We would like to express our appreciation to our students and Ms. Aastha Sharma at CRC (New Delhi) for her support and cooperation.

Ravi P. Agarwal
Cristina Flaut

Chapter 1

Linear Vector Spaces

A vector space (or linear space) consists of four things $\{F, V, +, \text{s.m.}\}$, where F is a field of scalars, V is the set of vectors, and $+$ and s.m. are binary operations on the set V called vector addition and scalar multiplication, respectively. In this chapter we shall define each term axiomatically and provide several examples.

Fields. A *field* is a set of scalars, denoted by F, in which two binary operations, addition $(+)$ and multiplication $(\cdot)$, are defined so that the following axioms hold:

A1. *Closure property of addition:* If $a, b \in F$, then $a + b \in F$.

A2. *Commutative property of addition:* If $a, b \in F$, then $a + b = b + a$.

A3. *Associative property of addition:* If $a, b, c \in F$, then $(a+b)+c = a+(b+c)$.

A4. *Additive identity:* There exists a zero element, denoted by 0, in F such that for all $a \in F$, $a + 0 = 0 + a = a$.

A5. *Additive inverse:* For each $a \in F$, there is a unique element $(-a) \in F$ such that $a + (-a) = (-a) + a = 0$.

A6. *Closure property of multiplication:* If $a, b \in F$, then $a \cdot b \in F$.

A7. *Commutative property of multiplication:* If $a, b \in F$, then $a \cdot b = b \cdot a$.

A8. *Associative property of multiplication:* If $a, b, c \in F$, then $(a \cdot b) \cdot c = a \cdot (b \cdot c)$.

A9. *Multiplicative identity:* There exists a unit element, denoted by 1, in F such that for all $a \in F$, $a \cdot 1 = 1 \cdot a = a$.

A10. *Multiplicative inverse:* For each $a \in F$, $a \neq 0$, there is an unique element $a^{-1} \in F$ such that $a \cdot a^{-1} = a^{-1}a = 1$.

A11. *Left distributivity:* If $a, b, c \in F$, then $a \cdot (b + c) = a \cdot b + a \cdot c$.

A12. *Right distributivity:* If $a, b, c \in F$, then $(a + b) \cdot c = a \cdot c + b \cdot c$.

Example 1.1. The set of rational numbers Q, the set of real numbers R, and the set of complex numbers C, with the usual definitions of addition and multiplication, are fields. The set of natural numbers $N = \{1, 2, \cdots\}$, and the set of all integers $Z = \{\cdots, -2, -1, 0, 1, 2 \cdots\}$ are not fields.

Let F and F_1 be fields and $F_1 \subseteq F$, then F_1 is called a *subfield* of F. Thus, Q is a subfield of R, and R is a subfield of C.

Vector spaces.   A *vector space* V over a field F denoted as (V, F) is a nonempty set of elements called *vectors* together with two binary operations, addition of vectors and multiplication of vectors by scalars, so that the following axioms hold:

B1. *Closure property of addition:* If $u, v \in V$, then $u + v \in V$.

B2. *Commutative property of addition:* If $u, v \in V$, then $u + v = v + u$.

B3. *Associativity property of addition:* If $u, v, w \in V$, then $(u + v) + w = u + (v + w)$.

B4. *Additive identity:* There exists a zero vector, denoted by 0, in V such that for all $u \in V$, $u + 0 = 0 + u = u$.

B5. *Additive inverse:* For each $u \in V$, there exists a vector v in V such that $u + v = v + u = 0$. Such a vector v is usually written as $-u$.

B6. *Closure property of multiplication:* If $u \in V$ and $a \in F$, then the product $a \cdot u = au \in V$.

B7. If $u, v \in V$ and $a \in F$, then $a(u + v) = au + av$.

B8. If $u \in V$ and $a, b \in F$, then $(a + b)u = au + bu$.

B9. If $u \in V$ and $a, b \in F$, then $ab(u) = a(bu)$.

B10. *Multiplication of a vector by a unit scalar:* If $u \in V$ and $1 \in F$, then $1u = u$.

In what follows, the subtraction of the vector v from u will be written as $u - v$, and by this we mean $u + (-v)$, or $u + (-1)v$. The spaces (V, R) and (V, C) will be called *real* and *complex vector spaces*, respectively.

Example 1.2 (The n-tuple space).   Let F be a given field. We consider the set V of all ordered n-tuples

$$
u \;=\; \begin{pmatrix} a_1 \\ \vdots \\ a_n \end{pmatrix} \qquad (\text{or,} \quad (a_1, \cdots, a_n))
$$

of scalars (known as *components*) $a_i \in F$. If

$$
v \;=\; \begin{pmatrix} b_1 \\ \vdots \\ b_n \end{pmatrix}
$$

is in V, the addition of u and v is defined by

$$
u + v \;=\; \begin{pmatrix} a_1 + b_1 \\ \vdots \\ a_n + b_n \end{pmatrix},
$$

and the product of a scalar $c \in F$ and vector $u \in V$ is defined by

$$cu = \begin{pmatrix} ca_1 \\ \vdots \\ ca_n \end{pmatrix}.$$

It is to be remembered that $u = v$, if and only if their corresponding components are equal, i.e., $a_i = b_i$, $i = 1, \cdots, n$. With this definition of addition and scalar multiplication it is easy to verify all the axioms B1–B10, and hence this (V, F) is a vector space. In particular, if

$$w = \begin{pmatrix} c_1 \\ \vdots \\ c_n \end{pmatrix}$$

is in V, then the i-th component of $(u + v) + w$ is $(a_i + b_i) + c_i$, which in view of A3 is the same as $a_i + (b_i + c_i)$, and this is the same as the i-th component of $u + (v + w)$, i.e., B3 holds. If $F = R$, then V is denoted as R^n, which for $n = 2$ and 3 reduces respectively to the two and three dimensional usual vector spaces. Similarly, if $F = C$, then V is written as C^n.

Example 1.3 (The space of polynomials). Let F be a given field. We consider the set $\mathcal{P}_n$, $n \geq 1$ of all polynomials of degree *at most* $n - 1$, i.e.,

$$\mathcal{P}_n = \left\{ a_0 + a_1 x + \cdots + a_{n-1} x^{n-1} = \sum_{i=0}^{n-1} a_i x^i : a_i \in F, \ x \in R \right\}.$$

If $u = \sum_{i=0}^{n-1} a_i x^i$, $v = \sum_{i=0}^{n-1} b_i x^i \in \mathcal{P}_n$, then the addition of vectors u and v is defined by

$$u + v = \sum_{i=0}^{n-1} a_i x^i + \sum_{i=0}^{n-1} b_i x^i = \sum_{i=0}^{n-1} (a_i + b_i) x^i,$$

and the product of a scalar $c \in F$ and vector $u \in \mathcal{P}_n$ is defined by

$$cu = c \sum_{i=0}^{n-1} a_i x^i = \sum_{i=0}^{n-1} (ca_i) x^i.$$

This $(\mathcal{P}_n, F)$ is a vector space. We remark that the set of all polynomials of degree exactly $n - 1$ is not a vector space. In fact, if we choose $b_{n-1} = -a_{n-1}$, then $u + v$ is a polynomial of degree $n - 2$.

Example 1.4 (The space of functions). Let F be a given field, and $X \subseteq F$. We consider the set V of all functions from the set X to F. The sum of two vectors $f, g \in V$ is defined by $(f + g)$, i.e., $(f + g)(x) = f(x) + g(x)$, $x \in X$, and the product of a scalar $c \in F$ and vector $f \in V$ is defined by cf, i.e., $(cf)(x) = cf(x)$. This (V, F) is a vector space. In particular, $(C[X], F)$, where $C[X]$ is the set of all continuous functions from X to F, with the same vector addition, and scalar multiplication is a vector space.

Example 1.5 (The space of sequences). Let F be a given field. Consider the set S of all sequences $a = \{a_n\}_{n=1}^{\infty}$, where $a_n \in F$. If a and b are in S and $c \in F$, we define $a + b = \{a_n\} + \{b_n\} = \{a_n + b_n\}$ and $ca = c\{a_n\} = \{ca_n\}$. Clearly, (S, F) is a vector space.

Example 1.6. Let $F = R$ and V be the set of all solutions of the homogeneous ordinary linear differential equation with real constant coefficients

$$a_0 \frac{d^n y}{dx^n} + a_1 \frac{d^{n-1} y}{dx^{n-1}} + \cdots + a_{n-1} \frac{dy}{dx} + a_n y = 0, \quad a_0 \neq 0, \quad x \in R.$$

This (V, F) is a vector space with the same vector addition and scalar multiplication as in Example 1.4. Note that if the above differential equation is nonhomogeneous then (V, F) is not a vector space.

Theorem 1.1. Let V be a vector space over the field F, and let $u, v \in V$. Then,

1. $u + v = u$ implies $v = 0 \in V$.
2. $0u = 0 \in V$.
3. $-u$ is unique.
4. $-u = (-1)u$.

Proof. 1. On adding $-u$ on both sides of $u + v = u$, we have

$$-u + u + v = -u + u \Rightarrow (-u + u) + v = 0 \Rightarrow 0 + v = 0 \Rightarrow v = 0.$$

2. Clearly, $0u = (0 + 0)u = 0u + 0u$, and hence $0u = 0 \in V$.

3. Assume that v and w are such that $u + v = 0$ and $u + w = 0$. Then, we have

$$v = v + 0 = v + (u + w) = (v + u) + w = (u + v) + w = 0 + w = w,$$

i.e., $-u$ of any vector $u \in V$ is unique.

4. Since

$$0 = 0u = [1 + (-1)]u = 1u + (-1)u = u + (-1)u,$$

it follows that $(-1)u$ is a negative for u. The uniqueness of this negative vector now follows from Part 3. ∎

Subspaces. Let (V, F) and (W, F) be vector spaces and $W \subseteq V$, then (W, F) is called a *subspace* of (V, F). It is clear that the smallest subspace

(W, F) of (V, F) consists of only the zero vector, and the largest subspace (W, F) is (V, F) itself.

Example 1.7. Let $F = R$,

$$W = \left\{ \begin{pmatrix} a_1 \\ a_2 \\ 0 \end{pmatrix} : a_1, a_2 \in R \right\} \quad \text{and} \quad V = \left\{ \begin{pmatrix} a_1 \\ a_2 \\ a_3 \end{pmatrix} : a_1, a_2, a_3 \in R \right\}.$$

Clearly, (W, R) is a subspace of (V, R). However, if we let

$$W = \left\{ \begin{pmatrix} a_1 \\ a_2 \\ a_3 \end{pmatrix} : a_1 > 0, a_2 > 0, a_3 > 0 \right\},$$

then (W, R) is not a subspace of (V, R).

Example 1.8. Let F be a given field. Consider the vector spaces $(\mathcal{P}_4, F)$ and $(\mathcal{P}_3, F)$. Clearly, $(\mathcal{P}_3, F)$ is a subspace of $(\mathcal{P}_4, F)$. However, the set of all polynomials of degree exactly two over the field F is not a subspace of $(\mathcal{P}_4, F)$.

Example 1.9. Consider the vector spaces (V, F) and $(C[X], F)$ considered in Example 1.4. Clearly, $(C[X], F)$ is a subspace of (V, F).

To check if the nonempty subset W of V over the field F is a subspace requires the verification of all the axioms B1–B10. However, the following result simplifies this verification considerably.

Theorem 1.2. If (V, F) is a vector space and W is a nonempty subset of V, then (W, F) is a subspace of (V, F) if and only if for each pair of vectors $u, v \in W$ and each scalar $a \in F$ the vector $au + v \in W$.

Proof. If (W, F) is a subspace of (V, F), and $u, v \in W$, $a \in F$, then obviously $au + v \in W$. Conversely, since $W \neq \emptyset$, there is a vector $u \in W$, and hence $(-1)u + u = 0 \in W$. Further, for any vector $u \in W$ and any scalar $a \in F$, the vector $au = au + 0 \in W$. This in particular implies that $(-1)u = -u \in W$. Finally, we notice that if $u, v \in W$, then $1u + v \in W$. The other axioms can be shown similarly. Thus (W, F) is a subspace of (V, F). ∎

Thus (W, F) is a subspace of (V, F) if and only if for each pair of vectors $u, v \in W, u + v \in W$ and for each scalar $a \in F$, $au \in W$.

Let $u^1, \cdots, u^n$ be vectors in a given vector space (V, F), and $c_1, \cdots, c_n \in F$ be scalars. The vector $u = c_1 u^1 + \cdots + c_n u^n$ is known as *linear combination* of u^i, $i = 1, \cdots, n$. By mathematical induction it follows that $u \in (V, F)$.

Theorem 1.3. Let $u^i \in (V, F)$, $i = 1, \cdots, n (\geq 1)$, and

$$W = \{c_1 u^1 + \cdots + c_n u^n : c_i \in F, \ i = 1, \cdots, n\}$$

then (W, F) is a subspace of (V, F), and W contains each of the vectors u^i, $i = 1, \cdots, n$.

Proof.   Clearly, each u^i is a linear combination of the form

$$u^i = \sum_{j=1}^{n} \delta_{ij} u^j,$$

where δ_{ij} is the Kronecker delta defined by

$$\delta_{ij} = \begin{cases} 0, & i \neq j \\ 1, & i = j. \end{cases}$$

Thus, each $u^i \in W$. Now, if $v = \sum_{i=1}^{n} c_i u^i$, $w = \sum_{i=1}^{n} d_i u^i$ and $a \in F$, then we have

$$av + w = a \sum_{i=1}^{n} c_i u^i + \sum_{i=1}^{n} d_i u^i = \sum_{i=1}^{n} (ac_i + d_i) u^i = \sum_{i=1}^{n} \alpha_i u^i, \quad \alpha_i \in F$$

which shows that $av + w \in W$. The result now follows from Theorem 1.2.   ∎

The subspace (W, F) in Theorem 1.3 is called the subspace *spanned* or *generated* by the vectors u^i, $i = 1, \cdots, n$, and written as $\text{Span}\{u^1, \cdots, u^n\}$. If $(W, F) = (V, F)$, then the set $\{u^1, \cdots, u^n\}$ is called a *spanning set* for the vector space (V, F). Clearly, in this case each vector $u \in V$ can be expressed as a linear combination of vectors u^i, $i = 1, \cdots, n$.

Example 1.10.   Since

$$2 \begin{pmatrix} 2 \\ 1 \\ 4 \end{pmatrix} - 3 \begin{pmatrix} 1 \\ 0 \\ 2 \end{pmatrix} + 5 \begin{pmatrix} 3 \\ 2 \\ 1 \end{pmatrix} - \begin{pmatrix} 4 \\ 2 \\ 0 \end{pmatrix} = \begin{pmatrix} 12 \\ 10 \\ 7 \end{pmatrix}$$

it follows that

$$\begin{pmatrix} 12 \\ 10 \\ 7 \end{pmatrix} \in \text{Span} \left\{ \begin{pmatrix} 2 \\ 1 \\ 4 \end{pmatrix}, \begin{pmatrix} 1 \\ 0 \\ 2 \end{pmatrix}, \begin{pmatrix} 3 \\ 2 \\ 1 \end{pmatrix}, \begin{pmatrix} 4 \\ 2 \\ 0 \end{pmatrix} \right\}.$$

However,

$$\begin{pmatrix} 1 \\ 2 \\ 3 \end{pmatrix} \notin \text{Span} \left\{ \begin{pmatrix} 1 \\ 0 \\ 0 \end{pmatrix}, \begin{pmatrix} 1 \\ 1 \\ 0 \end{pmatrix} \right\}.$$

Example 1.11.   For the vector space (V, F) considered in Example 1.2

the set $\{e^1, \cdots, e^n\}$, where

$$e^i = \begin{pmatrix} 0 \\ \vdots \\ 0 \\ 1 \\ 0 \\ \vdots \\ 0 \end{pmatrix} \in V \quad (1 \text{ at the } i\text{-th place})$$

is a spanning set. Similarly, for the vector space $(\mathcal{P}_n, F)$ considered in Example 1.3, the set $\{1, x, \cdots, x^{n-1}\}$ is a spanning set.

Problems

1.1. Show that the set of all real numbers of the form $a + \sqrt{2}b$, where a and b are rational numbers, is a field.

1.2. Show that

(i) if $u^1, \cdots, u^n$ span V and $u \in V$, then $u, u^1, \cdots, u^n$ also span V

(ii) if $u^1, \cdots, u^n$ span V and u^k is a linear combination of u^i, $i = 1, \cdots, n$, $i \neq k$, then u^i, $i = 1, \cdots, n$, $i \neq k$ also span V

(iii) if $u^1, \cdots, u^n$ span V and $u^k = 0$, then u^i, $i = 1, \cdots, n$, $i \neq k$ also span V.

1.3. Show that the intersection of any number of subspaces of a vector space V is a subspace of V.

1.4. Let U and W be subspaces of a vector space V. The space

$$U + W = \{v : v = u + w \text{ where } u \in U, \ w \in W\}$$

is called the *sum* of U and W. Show that

(i) $U + W$ is also a subspace of V

(ii) U and W are contained in $U + W$

(iii) $U + U = U$

(iv) $U \cup W$ is a subspace of V?.

1.5. Consider the following polynomials of degree three:

$$L_1(x) = \frac{(x - x_2)(x - x_3)(x - x_4)}{(x_1 - x_2)(x_1 - x_3)(x_1 - x_4)}, \quad L_2(x) = \frac{(x - x_1)(x - x_3)(x - x_4)}{(x_2 - x_1)(x_2 - x_3)(x_2 - x_4)}$$

$$L_3(x) = \frac{(x - x_1)(x - x_2)(x - x_4)}{(x_3 - x_1)(x_3 - x_2)(x_3 - x_4)}, \quad L_4(x) = \frac{(x - x_1)(x - x_2)(x - x_3)}{(x_4 - x_1)(x_4 - x_2)(x_4 - x_3)},$$

where $x_1 < x_2 < x_3 < x_4$. Show that

(i)   if $P_3(x) \in \mathcal{P}_4$ is an arbitrary polynomial of degree three, then $P_3(x) = L_1(x)P_3(x_1) + L_2(x)P_3(x_2) + L_3(x)P_3(x_3) + L_4(x)P_3(x_4)$

(ii)   the set $\{L_1(x), L_2(x), L_3(x), L_4(x)\}$ is a spanning set for $(\mathcal{P}_4, R)$.

1.6. Prove that the sets $\{1, 1+x, 1+x+x^2, 1+x+x^2+x^3\}$ and $\{1, (1-x), (1-x)^2, (1-x)^3\}$ are spanning sets for $(\mathcal{P}_4, R)$.

1.7. Let S be a subset of R^n consisting of all vectors with components a_i, $i = 1, \cdots, n$ such that $a_1 + \cdots + a_n = 0$. Show that S is a subspace of R^n.

1.8. On R^3 we define the following operations

$$\begin{pmatrix} x_1 \\ x_2 \\ x_3 \end{pmatrix} + \begin{pmatrix} y_1 \\ y_2 \\ y_3 \end{pmatrix} = \begin{pmatrix} x_1 + y_1 \\ 0 \\ x_3 + y_3 \end{pmatrix} \text{ and } a \begin{pmatrix} x_1 \\ x_2 \\ x_3 \end{pmatrix} = \begin{pmatrix} ax_1 \\ ax_2 \\ ax_3 \end{pmatrix}, \; a \in R.$$

With these operations, is R^3 a vector space over the field R?

1.9. Consider the following subsets of the vector space R^3:

(i) $V_1 = \{x \in R^3 : 3x_3 = x_1 - 5x_2\}$ (ii) $V_2 = \{x \in R^3 : x_1^2 = x_2 + 6x_3\}$

(iii) $V_3 = \{x \in R^3 : x_2 = 0\}$ (iv) $V_4 = \{x \in R^3 : x_2 = a, \; a \in R - \{0\}\}$.

Find if the above sets V_1, V_2, V_3, and V_4 are vector subspaces of R^3.

1.10. Let (V, X) be the vector space of functions considered in Example 1.4 with $X = F = R$, and $W \subset V$. Show that W is a subspace of V if

(i) W contains all bounded functions

(ii) W contains all even functions $(f(-x) = f(x))$

(iii) W contains all odd functions $(f(-x) = -f(x))$.

Answers or Hints

1.1. Verify A1–A12.

1.2. (i) Since $u^1, \cdots, u^n$ span V and $u \in V$ there exist scalars $c_1, \cdots, c_n$ such that $u = \sum_{i=1}^n c_i u^i$. Let $W = \{v : v = \sum_{i=1}^n \alpha_i u^i + \alpha_{n+1} u\}$. We need to show that $(V, F) = (W, F)$. Clearly, $V \subseteq W$. Now let $v \in W$, then $v = \sum_{i=1}^n \alpha_i u^i + \alpha_{n+1} \sum_{i=1}^n c_i u^i = \sum_{i=1}^n (\alpha_i + \alpha_{n+1} c_i) u^i$. Hence, $W \subseteq V$.

(ii) Similar as (i).

(iii) Similar as (i).

1.3. Let U, W be subspaces of V. It suffices to show that $U \cap W$ is also a subspace of V. Since $0 \in U$ and $0 \in W$ it is clear that $0 \in U \cap W$. Now let $u, w \in U \cap W$, then $u, w \in U$ and $u, w \in W$. Further for all scalars $a, b \in F$, $au + bw \in U$ and $au + bw \in W$. Thus $au + bw \in U \cap W$.

1.4. (i) Let $v^1, v^2 \in U + W$, where $v^1 = u^1 + w^1$, $v^2 = u^2 + w^2$. Then, $v^1 + v^2 = u^1 + w^1 + u^2 + w^2 = (u^1 + u^2) + (w^1 + w^2)$. Now since U and W are subspaces, $u^1 + u^2 \in U$ and $w^1 + w^2 \in W$. This implies that $v^1 + v^2 \in U + W$. Similarly we can show that $cv^1 \in U + W$, $c \in F$.

(ii) If $u \in U$, then since $0 \in W$, $u = u + 0 \in U + W$.

(iii) Since U is a subspace of V it is closed under vector addition, and hence $U + U \subseteq U$. We also have $U \subseteq U + U$ from (i).

(iv) $U \cup W$ need not be a subspace of V. For example, consider $V = R^3$,

$$U = \left\{ \begin{pmatrix} a_1 \\ 0 \\ 0 \end{pmatrix} : a_1 \in R \right\}, \quad W = \left\{ \begin{pmatrix} 0 \\ 0 \\ a_3 \end{pmatrix} : a_3 \in R \right\}.$$

Then

$$U \cup W = \left\{ \begin{pmatrix} a_1 \\ 0 \\ 0 \end{pmatrix}, \begin{pmatrix} 0 \\ 0 \\ a_3 \end{pmatrix} : a_1 \in R, \ a_3 \in R \right\}.$$

Clearly,

$$\begin{pmatrix} 1 \\ 0 \\ 0 \end{pmatrix} \in U \cup W, \quad \begin{pmatrix} 0 \\ 0 \\ 1 \end{pmatrix} \in U \cup W, \quad \text{but} \quad \begin{pmatrix} 1 \\ 0 \\ 1 \end{pmatrix} \notin U \cup W.$$

1.5. (i) The function $f(x) = L_1(x)P_3(x_1) + L_2(x)P_3(x_2) + L_3(x)P_3(x_3) + L_4(x)P_4(x_4)$ is a polynomial of degree at most three, and $f(x_i) = L_i(x_i) \times P_3(x_i) = P_3(x_i)$, $i = 1, 2, 3, 4$. Thus $f(x) = P_3(x)$ follows from the uniqueness of interpolating polynomials.

(ii) Follows from (i).

1.6. It suffices to note that $a + bx + cx^2 + dx^3 = (a - b) + (b - c)(1 + x) + (c - d)(1 + x + x^2) + d(1 + x + x^2 + x^3)$.

1.7. Use Theorem 1.2.

1.8. No.

1.9. V_1 and V_3 are vector subspaces, whereas V_2 and V_4 are not vector subspaces of R^3.

1.10. Use Theorem 1.2.

Chapter 2

Matrices

Matrices occur in many branches of applied mathematics and social sciences, such as algebraic and differential equations, mechanics, theory of electrical circuits, nuclear physics, aerodynamics, and astronomy. It is, therefore, necessary for every young scientist and engineer to learn the elements of matrix algebra.

A system of $m \times n$ elements from a field F arranged in a rectangular formation along m rows and n columns and bounded by the brackets ( ) is called an $m \times n$ *matrix*. Usually, a matrix is written by a single capital letter. Thus,

$$
A = \begin{pmatrix}
a_{11} & a_{12} & \cdots & a_{1j} & \cdots & a_{1n} \\
a_{21} & a_{22} & \cdots & a_{2j} & \cdots & a_{2n} \\
\cdots & \cdots & \cdots & \cdots & \cdots & \cdots \\
a_{i1} & a_{i2} & \cdots & a_{ij} & \cdots & a_{in} \\
\cdots & \cdots & \cdots & \cdots & \cdots & \cdots \\
a_{m1} & a_{m2} & \cdots & a_{mj} & \cdots & a_{mn}
\end{pmatrix}
$$

is an $m \times n$ matrix. In short, we often write $A = (a_{ij})$, where it is understood that the suffix $i = 1, \cdots, m$ and $j = 1, \cdots, n$, and ij indicates the i-th row and the j-th column. The numbers $(A)_{ij} = a_{ij}$ are called the *elements* of the matrix A. For example, the following matrices A and B are of order 2×3 and 3×2,

$$
A = \begin{pmatrix} 3 & 5 & 7 \\ 1 & 4 & 8 \end{pmatrix}, \quad B = \begin{pmatrix} 1+i & 1-i \\ 2+3i & 2-5i \\ 7 & 5+3i \end{pmatrix}, \quad i = \sqrt{-1}.
$$

A matrix having a single row, i.e., $m = 1$, is called a *row matrix* or a *row vector*, e.g., $(2 \ 3 \ 5 \ 7)$.

A matrix having a single column, i.e., $n = 1$, is called a *column matrix* or a *column vector*, e.g.,

$$
\begin{pmatrix} 5 \\ 7 \\ 3 \end{pmatrix}.
$$

Thus the columns of the matrix A can be viewed as vertical m-tuples (see

Example 1.2), and the rows as horizontal n-tuples. Hence, if we let

$$a^j = \begin{pmatrix} a_{1j} \\ \vdots \\ a_{mj} \end{pmatrix}, \quad j = 1, 2, \cdots, n$$

then the above matrix A can be written as

$$A = (a^1, a^2, \cdots, a^n).$$

A matrix having n rows and n columns is called a *square matrix* of order n, e.g.,

$$A = \begin{pmatrix} 1 & 2 & 3 \\ 2 & 3 & 4 \\ 3 & 4 & 5 \end{pmatrix} \tag{2.1}$$

is a square matrix of order 3.

For a square matrix A of order n, the elements a_{ii}, $i = 1, \cdots, n$, lying on the *leading* or *principal diagonal* are called the *diagonal elements* of A, whereas the remaining elements are called the *off-diagonal* elements. Thus for the matrix A in (2.1) the diagonal elements are $1, 3, 5$.

A square matrix all of whose elements except those in the principal diagonal are zero, i.e., $a_{ij} = 0$, $|i - j| \geq 1$ is called a *diagonal matrix*, e.g.,

$$A = \begin{pmatrix} 7 & 0 & 0 \\ 0 & 5 & 0 \\ 0 & 0 & 1 \end{pmatrix}.$$

A diagonal matrix of order n that has unity for all its diagonal elements, i.e., $a_{ii} = 1$, is called a *unit* or *identity matrix* of order n and is denoted by I_n or simply by I. For example, identity matrix of order 3 is

$$I_3 = \begin{pmatrix} 1 & 0 & 0 \\ 0 & 1 & 0 \\ 0 & 0 & 1 \end{pmatrix},$$

and of nth order $I_n = (e^1, e^2, \cdots, e^n)$.

If all the elements of a matrix are zero, i.e., $a_{ij} = 0$, it is called a *null* or *zero matrix* and is denoted by 0, e.g.,

$$0 = \begin{pmatrix} 0 & 0 \\ 0 & 0 \\ 0 & 0 \end{pmatrix}.$$

A square matrix $A = (a_{ij})$ is called *symmetric* when $a_{ij} = a_{ji}$. If $a_{ij} = -a_{ji}$, so that all the principal diagonal elements are zero, then the matrix is called a *skew-symmetric matrix*. Examples of symmetric and skew-symmetric matrices are respectively

$$
\begin{pmatrix} a & h & g \\ h & b & f \\ g & f & c \end{pmatrix}
\quad \text{and} \quad
\begin{pmatrix} 0 & h & -g \\ -h & 0 & f \\ g & -f & 0 \end{pmatrix}.
$$

An $m \times n$ matrix is called *upper triangular* if $a_{ij} = 0$, $i > j$ and *lower triangular* if $a_{ij} = 0$, $j > i$. In particular, a square matrix all of whose elements below the principal diagonal are zero is called an *upper triangular matrix*, and a square matrix all of whose elements above the principal diagonal are zero is called a *lower triangular matrix*. Thus,

$$
\begin{pmatrix} a & h & g \\ 0 & b & f \\ 0 & 0 & c \end{pmatrix}
\quad \text{and} \quad
\begin{pmatrix} a & 0 & 0 \\ h & b & 0 \\ g & f & c \end{pmatrix}
$$

are upper and lower triangular matrices, respectively. Clearly, a square matrix is diagonal if and only if it is both upper and lower triangular.

Two matrices $A = (a_{ij})$ and $B = (b_{ij})$ are said to be *equal* if and only if they are of the same order, and $a_{ij} = b_{ij}$ for all i and j.

If A and B are two matrices of the same order, then their sum $A + B$ is defined as the matrix each element of which is the sum of the corresponding elements of A and B. Thus,

$$
\begin{pmatrix} a_1 & b_1 \\ a_2 & b_2 \\ a_3 & b_3 \end{pmatrix}
+
\begin{pmatrix} c_1 & d_1 \\ c_2 & d_2 \\ c_3 & d_3 \end{pmatrix}
=
\begin{pmatrix} a_1 + c_1 & b_1 + d_1 \\ a_2 + c_2 & b_2 + d_2 \\ a_3 + c_3 & b_3 + d_3 \end{pmatrix}.
$$

Similarly, $A - B$ is defined as a matrix whose elements are obtained by subtracting the elements of B from the corresponding elements of A. Thus,

$$
\begin{pmatrix} a_1 & b_1 \\ a_2 & b_2 \end{pmatrix}
-
\begin{pmatrix} c_1 & d_1 \\ c_2 & d_2 \end{pmatrix}
=
\begin{pmatrix} a_1 - c_1 & b_1 - d_1 \\ a_2 - c_2 & b_2 - d_2 \end{pmatrix}.
$$

The addition of matrices satisfies the following properties:

1. $A + B = B + A$, *commutative law*
2. $A + (B + C) = (A + B) + C$, *associative law*
3. $A + 0 = 0 + A = A$,
4. $A + (-A) = (-A) + A = 0$.

The product of a matrix A by a scalar $k \in F$ is a matrix whose every element is k times the corresponding element of A. Thus,

$$
k \begin{pmatrix} a_1 & b_1 & c_1 \\ a_2 & b_2 & c_2 \end{pmatrix}
=
\begin{pmatrix} ka_1 & kb_1 & kc_1 \\ ka_2 & kb_2 & kc_2 \end{pmatrix}.
$$

For such products, the following properties hold:

1. $(k_1 + k_2)A = k_1 A + k_2 A, \ k_1, k_2 \in F$
2. $k_1(k_2 A) = (k_1 k_2)A$
3. $k(A + B) = kA + kB$, *distributive law*
4. $(-1)A = -A$
5. $0A = 0$
6. $k0 = 0$.

In what follows we shall denote by $M^{m \times n}$ the set of all $m \times n$ matrices whose elements belong to a certain field F, i.e.,

$$M^{m \times n} = \{A = (a_{ij}) : a_{ij} \in F, \ i = 1, \cdots, m, \ n = 1, 2, \cdots, n\}.$$

It is clear that with the above definition of addition and scalar multiplication, $(M^{m \times n}, F)$ is a vector space. The set of all $m \times n$ matrices with real (complex) elements will be represented by $R^{m \times n}$ ($C^{m \times n}$).

Two matrices can be multiplied only when the number of columns in the first matrix is equal to the number of rows in the second matrix. Such matrices are said to be *conformable* for multiplication. Thus, if A and B are $n \times m$ and $m \times p$ matrices

$$A = \begin{pmatrix} a_{11} & a_{12} & \cdots & a_{1m} \\ a_{21} & a_{22} & \cdots & a_{2m} \\ \cdots & \cdots & \cdots & \cdots \\ a_{n1} & a_{n2} & \cdots & a_{nm} \end{pmatrix} \quad \text{and} \quad B = \begin{pmatrix} b_{11} & b_{12} & \cdots & b_{1p} \\ b_{21} & b_{22} & \cdots & b_{2p} \\ \cdots & \cdots & \cdots & \cdots \\ a_{m1} & a_{m2} & \cdots & a_{mp} \end{pmatrix}$$

then $A \times B$, or simply AB is a new matrix of order $n \times p$,

$$AB = \begin{pmatrix} c_{11} & c_{12} & \cdots & c_{1p} \\ c_{21} & c_{22} & \cdots & c_{2p} \\ \cdots & \cdots & \cdots & \cdots \\ c_{n1} & c_{n2} & \cdots & c_{np} \end{pmatrix},$$

where

$$c_{ij} = a_{i1}b_{1j} + \cdots + a_{im}b_{mj} = \sum_{k=1}^{m} a_{ik}b_{kj}.$$

Thus, in particular

$$\begin{pmatrix} a_1 & b_1 & c_1 \\ a_2 & b_2 & c_2 \\ a_3 & b_3 & c_3 \end{pmatrix} \times \begin{pmatrix} d_1 & e_1 \\ d_2 & e_2 \\ d_3 & e_3 \end{pmatrix} = \begin{pmatrix} a_1 d_1 + b_1 d_2 + c_1 d_3 & a_1 e_1 + b_1 e_2 + c_1 e_3 \\ a_2 d_1 + b_2 d_2 + c_2 d_3 & a_2 e_1 + b_2 e_2 + c_2 e_3 \\ a_3 d_1 + b_3 d_2 + c_3 d_3 & a_3 e_1 + b_3 e_2 + c_3 e_3 \end{pmatrix}.$$

In the product AB the matrix A is said to be *post-multiplied* by B, and the matrix B is said to be *pre-multiplied* by A. It is possible that AB is defined, but BA may not be defined. Further, both AB and BA may exist yet may not be equal.

Example 2.1. For the matrices

$$A = \begin{pmatrix} 0 & 1 & 2 \\ 1 & 2 & 3 \\ 2 & 3 & 4 \end{pmatrix}, \quad B = \begin{pmatrix} 1 & -2 \\ -1 & 0 \\ 2 & -1 \end{pmatrix},$$

we have

$$AB = \begin{pmatrix} 3 & -2 \\ 5 & -5 \\ 7 & -8 \end{pmatrix}.$$

However, BA is not defined.

Example 2.2. For the matrices

$$A = \begin{pmatrix} 1 & 1 & 0 \\ -1 & 2 & 1 \\ 0 & 0 & 2 \end{pmatrix}, \quad B = \begin{pmatrix} 2 & 3 & 4 \\ 1 & 2 & 3 \\ -1 & 1 & 2 \end{pmatrix},$$

we have

$$AB = \begin{pmatrix} 3 & 5 & 7 \\ -1 & 2 & 4 \\ -2 & 2 & 4 \end{pmatrix}, \quad BA = \begin{pmatrix} -1 & 8 & 11 \\ -1 & 5 & 8 \\ -2 & 1 & 5 \end{pmatrix}.$$

Thus, $AB \neq BA$.

Example 2.3. For the matrices

$$A = \begin{pmatrix} 1 & 1 \\ 1 & 1 \end{pmatrix}, \quad B = \begin{pmatrix} 1 & -1 \\ -1 & 1 \end{pmatrix},$$

we have $AB = 0$. Thus $AB = 0$ does not imply that A or B is a null matrix.

For the multiplication of matrices, the following properties hold:

1. $A(BC) = (AB)C$, *associative law*
2. $A(B \pm C) = AB \pm AC$ and $(A \pm B)C = AC \pm BC$, *distributive law*
3. $AI = A$ and $IA = A$
4. $k(AB) = (kA)B = A(kB)$, $k \in F$
5. $A0 = 0$ and $0B = 0$.

If A is an $n \times n$ matrix, then the product AA is denoted as A^2. In general, the m times product $AA \cdots A = A^{m-1}A = A^m$ and $A^0 = I$. Also, if m and p are positive integers, then in view of the associative law, we have $A^m A^p = A^{m+p}$. In particular, $I = I^2 = I^3 = \cdots$. Further, if A is a diagonal matrix with diagonal elements $(\lambda_1, \cdots, \lambda_n)$, then A^m is also a diagonal matrix with diagonal elements $(\lambda_1^m, \cdots, \lambda_n^m)$. Moreover, if A is an upper (lower) triangular matrix, then A^m is also an upper (lower) triangular matrix. Polynomials in the matrix A are also defined. In fact, if $P_{m-1}(x) = \sum_{i=0}^{m-1} a_i x^i \in (\mathcal{P}_m, F)$, then $P_{m-1}(A) = \sum_{i=0}^{m-1} a_i A^i$. Clearly, $P_{m-1}(A)$ is a square matrix. If $P_{m-1}(A)$ is the zero matrix, then A is called a *zero* or *root* of $P_{m-1}(x)$.

The *transpose* of an $m \times n$ matrix $A = (a_{ij})$, written as A^t, is an $n \times m$ matrix that is obtained by interchanging the rows and columns of A, i.e., $A^t = (a_{ji})$. Thus for the matrices in Example 2.1,

$$A^t = \begin{pmatrix} 0 & 1 & 2 \\ 1 & 2 & 3 \\ 2 & 3 & 4 \end{pmatrix}, \quad B^t = \begin{pmatrix} 1 & -1 & 2 \\ -2 & 0 & -1 \end{pmatrix}.$$

It follows that for the column vector

$$a = \begin{pmatrix} a_1 \\ \vdots \\ a_n \end{pmatrix}$$

a^t is the row vector $a^t = (a_1, \cdots, a_n)$, and vice-versa.

For the transpose of matrices, the following hold:

1. $(A + B)^t = A^t + B^t$
2. $(cA)^t = cA^t$, where c is a scalar
3. $(A^t)^t = A$
4. $(AB)^t = B^t A^t$ (note reversed order).

Clearly, a square matrix A is symmetric if and only if $A = A^t$ and skew-symmetric if and only if $A = -A^t$. If A is symmetric (skew-symmetric), then obviously A^t is symmetric (skew-symmetric).

The *trace* of a square matrix, written as $\text{tr}(A)$, is the sum of the diagonal elements, i.e., $\text{tr}(A) = a_{11} + a_{22} + \cdots + a_{nn}$. Thus for the matrix A in (2.1) the trace is $1 + 3 + 5 = 9$. For the trace of square matrices, the following hold:

1. $\text{tr}(A + B) = \text{tr}(A) + \text{tr}(B)$
2. $\text{tr}(A) = \text{tr}(A^t)$
3. $\text{tr}(cA) = c \, \text{tr}(A)$, where c is a scalar
4. $\text{tr}(AB) = \text{tr}(BA)$.

Finally, we remark that, especially for computational purposes, a matrix A can be *partitioned* into submatrices called *blocks*, or *cells* by drawing horizontal and vertical lines between its rows and columns. This partition is not unique;

for example, the matrix

$$A = \begin{pmatrix} 1 & 0 & 7 & 3 & 4 \\ 3 & 5 & 7 & 1 & 0 \\ 0 & 4 & 3 & 2 & 7 \\ 6 & 3 & 9 & 0 & 2 \end{pmatrix}$$

can be partitioned as

$$\left(\begin{array}{ccccc} 1 & 0 & 7 & 3 & 4 \\ 3 & 5 & 7 & 1 & 0 \\ \hline 0 & 4 & 3 & 2 & 7 \\ 6 & 3 & 9 & 0 & 2 \end{array} \right) = \begin{pmatrix} A_{11} & A_{12} \\ A_{21} & A_{22} \end{pmatrix},$$

or

$$\left(\begin{array}{cc|c|cc} 1 & 0 & 7 & 3 & 4 \\ \hline 3 & 5 & 7 & 1 & 0 \\ 0 & 4 & 3 & 2 & 7 \\ \hline 6 & 3 & 9 & 0 & 2 \end{array} \right) = \begin{pmatrix} A_{11} & A_{12} & A_{13} \\ A_{21} & A_{22} & A_{23} \\ A_{31} & A_{32} & A_{33} \end{pmatrix}.$$

If $A = (A_{ij})_{r \times s}$, $B = (B_{ij})_{s \times t}$ and the blocks A_{ij}, B_{ij} are conformable, then $AB = (C_{ij})_{r \times t}$, where

$$C_{ij} = \sum_{k=1}^{s} A_{ik} B_{kj}.$$

In partitioned matrices the blocks can be treated as numbers so that the basic operations between matrices (with blocks of correct orders) can be performed.

Problems

2.1. Express the following matrix as a sum of a lower triangular matrix and an upper triangular matrix with zero leading diagonal

$$A = \begin{pmatrix} 1 & 3 & 5 \\ 2 & 4 & 7 \\ 6 & -5 & 9 \end{pmatrix}.$$

2.2. Let $A = (a_{ij}), B = (b_{ij}) \in M^{n \times n}$ be upper triangular matrices. Show that AB is an upper triangular matrix with diagonal elements $a_{ii}b_{ii}$, $i = 1, \cdots, n$.

2.3. For the matrices

$$A = \begin{pmatrix} -1 & 2 & 4 \\ 3 & 6 & 5 \\ 6 & 5 & 8 \end{pmatrix}, \quad B = \begin{pmatrix} 3 & -6 & -5 \\ 1 & 3 & 5 \\ 3 & 5 & 7 \end{pmatrix}$$

find $A+B$, $A-B$, $2A+3B$, $3A-4B$, AB, BA, A^2 and B^3.

2.4. For the matrices

$$A = \begin{pmatrix} 2 & -3 & -5 \\ -1 & 4 & 5 \\ 1 & -3 & -4 \end{pmatrix}, \quad B = \begin{pmatrix} -1 & 3 & 5 \\ 1 & -3 & -5 \\ -1 & 3 & 5 \end{pmatrix}, \quad C = \begin{pmatrix} 2 & -2 & -4 \\ -1 & 3 & 4 \\ 1 & -3 & -4 \end{pmatrix}$$

verify that $AB = BA = 0$, $AC \neq A$, and $CA = C$.

2.5. If

$$A = \begin{pmatrix} \cos\theta & \sin\theta \\ -\sin\theta & \cos\theta \end{pmatrix}$$

show that

$$A^n = \begin{pmatrix} \cos n\theta & \sin n\theta \\ -\sin n\theta & \cos n\theta \end{pmatrix}.$$

2.6. Show that $(A+B)^2 = A^2+2AB+B^2$ and $(A+B)(A-B) = A^2-B^2$ if and only if the square matrices A and B commute.

2.7. Consider the set $R^{2\times 2}$ with the addition as usual, but the scalar multiplication as follows:

$$k \begin{pmatrix} a & b \\ c & d \end{pmatrix} = \begin{pmatrix} ka & 0 \\ 0 & kd \end{pmatrix}.$$

Show that $(R^{2\times 2}, R)$ is not a vector space.

2.8. Show that the matrices

$$\begin{pmatrix} 1 & 0 \\ 0 & 0 \end{pmatrix}, \quad \begin{pmatrix} 0 & 1 \\ 0 & 0 \end{pmatrix}, \quad \begin{pmatrix} 0 & 0 \\ 1 & 0 \end{pmatrix}, \quad \begin{pmatrix} 0 & 0 \\ 0 & 1 \end{pmatrix}$$

span the vector space $M^{2\times 2}$ containing all 2×2 matrices.

2.9. Let $B \in R^{n\times n}$ be a fixed matrix, and $S = \{A : AB = BA, A \in R^{n\times n}\}$. Show that S is a subspace of $R^{n\times n}$.

2.10. Let $A_l \in M^{m\times n}$, $k_l \in F$, $l = 1, 2, \cdots, M$. Show that

(i) $\quad k_1 \sum_{l=1}^{M} A_l = \sum_{l=1}^{M} k_1 A_l$

(ii) $\quad \left(\sum_{l=1}^{M} k_l \right) A_1 = \sum_{l=1}^{M} k_l A_1.$

2.11. For the matrix multiplication, prove associative and distributive laws.

2.12. For the transpose of matrices, show that $(AB)^t = B^t A^t$.

2.13. The *hermitian transpose* of a complex $m \times n$ matrix $A = (a_{ij})$, written as A^H, is an $n \times m$ matrix that is obtained by interchanging the rows and columns of A and taking the complex conjugate of the elements (if $z = a + ib$, then $\bar{z} = a - ib$ is its complex conjugate), i.e., $A^H = (\bar{a}_{ji})$. For the hermitian transpose of matrices, show that

(i) $(A + B)^H = A^H + B^H$

(ii) $(cA)^H = \bar{c}A^H$, where c is a scalar

(iii) $(A^H)^H = A$

(iv) $(AB)^H = B^H A^H$.

2.14. A square complex matrix A is called *hermitian* if and only if $A = A^H$, *skew-hermitian* if and only if $A = -A^H$, and *normal* if and only if A commutes with A^H, i.e., $AA^H = A^H A$. Give some examples of hermitian, skew-hermitian, and normal matrices.

2.15. Show that

(i)   the addition $A + B$ of two symmetric (hermitian) matrices A and B is symmetric (hermitian), but the product AB is symmetric (hermitian) if and only if A and B commute, in particular AA^t and $A^t A$ (AA^H and $A^H A$) are symmetric (hermitian)

(ii)   if A is an $n \times n$ symmetric (hermitian) matrix and B is any $n \times m$ matrix, then $B^t AB$ ($B^H AB$) is symmetric (hermitian)

(iii)   if A is a symmetric (hermitian) matrix, then for all positive integers p, A^p is symmetric (hermitian)

(iv)   if A^2 is a symmetric (hermitian) matrix, then A need not be symmetric (hermitian)

(v)   a skew–symmetric (skew-hermitian) matrix must be square and its diagonal elements must be zero

(vi)   for a given square matrix A the matrix $A - A^t$ ($A - A^H$) is skew–symmetric (skew-hermitian) while the matrix $A + A^t$ ($A + A^H$) is symmetric (hermitian)

(vii)   any square matrix can be uniquely written as the sum of a symmetric (hermitian) and a skew–symmetric (skew-hermitian) matrix

(viii) if A is a skew–symmetric (skew-hermitian) $n \times n$ matrix, then for any $u \in R^n$ (C^n), uAu^t (uAu^H) $= 0$.

2.16. Give an example of two matrices $A, B \in C^{n \times n}$ for which $AB \neq BA$ but $\text{tr}(AB) = \text{tr}(BA)$, and hence deduce that $AB - BA = I$ cannot be valid. Further, show that $\text{tr}(A^H A) \geq 0$.

2.17. A real $n \times n$ matrix that has nonnegative elements and where each column adds up to 1 is called a *stochastic matrix*. If a stochastic matrix also has rows that add to 1, then it is called a *doubly stochastic matrix*. Show that if A and B are $n \times n$ stochastic matrices, then AB is also an stochastic matrix.

Answers or Hints

2.1. $U = \begin{pmatrix} 0 & 3 & 5 \\ 0 & 0 & 7 \\ 0 & 0 & 0 \end{pmatrix}$, $L = \begin{pmatrix} 1 & 0 & 0 \\ 2 & 4 & 0 \\ 6 & -5 & 9 \end{pmatrix}$.

2.2. $AB = C = (c_{ij})$, $c_{ij} = 0$ if $i > j$ and $c_{ij} = \sum_{k=i}^{j} a_{ik} b_{kj}$ if $j \geq i$.

2.3. Direct computation.

2.4. Direct computation.

2.5. Direct computation.

2.6. $(A+B)^2 = (A+B)(A+B) = A^2 + AB + BA + B^2$.

2.7. Consider $A = \begin{pmatrix} 1 & 1 \\ 1 & 1 \end{pmatrix}$. Then $1 \cdot A = \begin{pmatrix} 1 & 0 \\ 0 & 1 \end{pmatrix} \neq A$. Hence, B10 is violated.

2.8. $\begin{pmatrix} a & b \\ c & d \end{pmatrix} = \begin{pmatrix} a & 0 \\ 0 & 0 \end{pmatrix} + \begin{pmatrix} 0 & b \\ 0 & 0 \end{pmatrix} + \begin{pmatrix} 0 & 0 \\ c & 0 \end{pmatrix} + \begin{pmatrix} 0 & 0 \\ 0 & d \end{pmatrix}$.

2.9. Let $C, D \in S$ and $\alpha, \beta \in R$. Then, $(\alpha C + \beta D)B = \alpha CB + \beta DB = \alpha BC + \beta BD = B(\alpha C + \beta D)$.

2.10. Use the principal of mathematical induction.

2.11. Let $A = (a_{ij})_{m \times n}$, $B = (b_{ij})_{n \times p}$, $C = (c_{ij})_{p \times r}$, then BC and AB are $n \times r$ and $m \times p$ matrices, and the ij-th element of $A(BC)$ is

$$\sum_{\mu=1}^{n} a_{i\mu} \left(\sum_{\nu=1}^{p} b_{\mu\nu} c_{\nu j} \right) = \sum_{\mu=1}^{n} \sum_{\nu=1}^{p} a_{i\mu} b_{\mu\nu} c_{\nu j}$$

and similarly, the ij-th element of $(AB)C$ is

$$\sum_{\nu=1}^{p} \left(\sum_{\mu=1}^{n} a_{i\mu} b_{\mu\nu} \right) c_{\nu j} = \sum_{\mu=1}^{n} \sum_{\nu=1}^{p} a_{i\mu} b_{\mu\nu} c_{\nu j}.$$

2.12. Similar to Problem 2.11.

2.13. (iv) $(AB)_{ij}^{H} = (\sum_{k=1}^{n} a_{ik} b_{kj})_{ij}^{H} = (\overline{\sum_{k=1}^{n} a_{ik} b_{kj}})_{ji} = (\sum_{k=1}^{n} \bar{a}_{ik} \bar{b}_{kj})_{ji}$ $= (\overline{AB})_{ij}^{t} = (\bar{B}^{t} \bar{A}^{t})_{ij} = (B^{H} A^{H})_{ij}$.

2.14. Hermitian: $\begin{pmatrix} 1 & i & 2-i \\ -i & 2 & 5 \\ 2+i & 5 & 3 \end{pmatrix}$

Skew-hermitian: $\begin{pmatrix} i & i & 2+i \\ i & 2i & 5 \\ -2+i & -5 & 5i \end{pmatrix}$

Normal: $\begin{pmatrix} -i & -2-i \\ 2-i & -3i \end{pmatrix}$.

2.15. (i) Let AB be hermitian. Then, $AB = (AB)^{H} = B^{H} A^{H} = BA$, i.e., A, B commute. Conversely, let A, B commute. Then, $AB = BA = B^{H} A^{H} = (AB)^{H}$, i.e., AB is hermitian.

(vii) $A = \frac{1}{2}(A - A^H) + \frac{1}{2}(A + A^H)$.

2.16. Let $A = \begin{pmatrix} 1+i & 2 \\ 1 & -i \end{pmatrix}$, $B = \begin{pmatrix} 2i & 1 \\ 1-i & 3 \end{pmatrix}$. $\operatorname{tr}(AB) - \operatorname{tr}(BA) = 0 \neq$ $n = \operatorname{tr}(I)$. $\operatorname{tr}(A^H A) = \sum_{i=1}^{n} \sum_{j=1}^{n} a_{ij} \bar{a}_{ij} \geq 0$.

2.17. Check for $n = 2$, and see the pattern.

Chapter 3

Determinants

Many complicated expressions, particularly in electrical and mechanical engineering, can be elegantly solved by expressing them in the form of determinants. Further, determinants of orders 2 and 3 geometrically represent areas and volumes, respectively. Therefore, the working knowledge of determinants is a basic necessity for all science and engineering students. In this chapter, we shall briefly sketch the important properties of determinants. The applications of determinants to find the solutions of linear systems of algebraic equations will be presented in Chapter 6.

Associated with a square $n \times n$ matrix $A = (a_{ij}) \in M^{n \times n}$ there is a scalar in F called the *determinant* of order n of A, and it is denoted as $\det(A)$, or $|A|$, or

$$\begin{vmatrix} a_{11} & a_{12} & \cdots & a_{1n} \\ a_{21} & a_{22} & \cdots & a_{2n} \\ \cdots & & & \\ a_{n1} & a_{n2} & \cdots & a_{nn} \end{vmatrix}.$$

The determinants of orders 1 and 2 are defined as

$$|a_{11}| = a_{11}, \qquad \begin{vmatrix} a_{11} & a_{12} \\ a_{21} & a_{22} \end{vmatrix} = a_{11}a_{22} - a_{12}a_{21}.$$

If in the matrix A we choose any p rows and any p columns, where $p \leq n$, then the elements at the intersection of these rows and columns form a square matrix of order p. The determinant of this new matrix is called a *minor of pth order* of the matrix A. A minor of any diagonal element of A is called a *principal minor*. In particular, an $(n-1) \times (n-1)$ determinant obtained by deleting i-th row and j-th column of the matrix A is the *minor of* $(n-1)th$ *order*, which we denote as $\tilde{a}_{ij}$, and call $\alpha_{ij} = (-1)^{i+j}\tilde{a}_{ij}$ the *cofactor* of a_{ij}. In terms of cofactors the determinant of A is defined as

$$|A| = \sum_{j=1}^{n} a_{ij}\alpha_{ij} = \sum_{i=1}^{n} a_{ij}\alpha_{ij}. \tag{3.1}$$

Further,

$$\sum_{j=1}^{n} a_{ij}\alpha_{kj} = 0 \quad \text{if} \quad i \neq k \tag{3.2}$$

and

$$\sum_{i=1}^{n} a_{ij}\alpha_{ik} = 0 \quad \text{if} \quad j \neq k. \tag{3.3}$$

Thus a determinant of order n can be written in terms of n determinants of order $n-1$. Formula (3.1) for computing the determinant of A is called the *Laplace expansion*. In particular, for the determinant of order 3, we have

$$
\begin{vmatrix} a_{11} & a_{12} & a_{13} \\ a_{21} & a_{22} & a_{23} \\ a_{31} & a_{32} & a_{33} \end{vmatrix} = a_{11} \begin{vmatrix} a_{22} & a_{23} \\ a_{32} & a_{33} \end{vmatrix} - a_{12} \begin{vmatrix} a_{21} & a_{23} \\ a_{31} & a_{33} \end{vmatrix} + a_{13} \begin{vmatrix} a_{21} & a_{22} \\ a_{31} & a_{32} \end{vmatrix}
$$

$$= a_{11}(a_{22}a_{33} - a_{23}a_{32}) - a_{12}(a_{21}a_{33} - a_{23}a_{31})$$

$$+ a_{13}(a_{21}a_{32} - a_{22}a_{31})$$

$$= a_{11}a_{22}a_{33} - a_{11}a_{23}a_{32} - a_{12}a_{21}a_{33} + a_{12}a_{23}a_{31}$$

$$+ a_{13}a_{21}a_{32} - a_{13}a_{22}a_{31}. \tag{3.4}$$

To find a general expression for the determinants of order n similar to (3.4), we recall that a permutation σ of the set $N = \{1, 2, \cdots, n\}$ is a one-to-one mapping of N into itself. Such a permutation is generally denoted as $\sigma = i_1 i_2 \cdots i_n$, where $i_j = \sigma(j)$. It is clear that there are $n!$ permutations of N. The set of all such permutations is denoted as S_n. As an example, for the set $\{1, 2, 3\}$ there are 6 permutations, and $S_3 = \{123, 132, 213, 231, 312, 321\}$. If $\sigma \in S_n$, then the inverse mapping $\sigma^{-1} \in S_n$, and if $\sigma, \tau \in S_n$, then the composite mapping $\sigma \circ \tau \in S_n$. Further, the identity mapping $\sigma \circ \sigma^{-1} = 12 \cdots n \in S_n$. By an *inversion* in σ we mean a pair of integers (i, j) such that $i > j$, but i precedes j in σ. We say σ is an even or odd permutation according to whether there is an even or odd number of inversions in σ. We also define

$$\text{sgn } \sigma = \begin{cases} 1 & \text{if } \sigma \text{ has even permutation} \\ -1 & \text{if } \sigma \text{ has odd permutation}. \end{cases}$$

Equivalently, we can define a permutation to be even or odd in accordance with whether the minimum number of interchanges required to put the permutation in natural order is even or odd. It can be shown that for any n, half of the permutations in S_n are even and half of them are odd.

With this terminology it follows that

$$|A| = \sum_{\sigma \in S_n} (\text{sgn } \sigma) a_{1\sigma(1)} a_{2\sigma(2)} \cdots a_{n\sigma(n)}. \tag{3.5}$$

Example 3.1. In $12 \cdots n$ the inversion is zero (even), whereas in 321 the inversion is three (odd) because there are two numbers (3 and 2) greater than and preceding 1, and one number (3) greater than and preceding 2. In the set S_3 the permutations $123, 231, 312$ are even, and $321, 213, 132$ are odd, thus from (3.5) the expansion (3.4) follows. In $4312 \in S_4$, 4 precedes $3, 1$ and 2, and 3 precedes 1 and 2; thus the inversion is 5 (odd).

Example 3.2. From (3.5) it immediately follows that for the lower triangular and upper triangular matrices A, we have

$$|A| = \begin{vmatrix} a_{11} & 0 & \cdots & 0 \\ a_{21} & a_{22} & \cdots & 0 \\ \cdots & & & \\ a_{n1} & a_{n2} & \cdots & a_{nn} \end{vmatrix} = a_{11}a_{22}\cdots a_{nn}$$

and

$$|A| = \begin{vmatrix} a_{11} & a_{12} & \cdots & a_{1n} \\ 0 & a_{22} & \cdots & a_{2n} \\ \cdots & & & \\ 0 & 0 & \cdots & a_{nn} \end{vmatrix} = a_{11}a_{22}\cdots a_{nn}.$$

Thus, in particular, for the identity matrix $|I| = 1$. Similarly, it follows that

$$\begin{vmatrix} 0 & 0 & \cdots & 0 & a_{1n} \\ 0 & 0 & \cdots & a_{2,n-1} & a_{2n} \\ \cdots & & & & \\ 0 & a_{n-1,2} & \cdots & a_{n-1,n-1} & a_{n-1,n} \\ a_{n1} & a_{n2} & \cdots & a_{n,n-1} & a_{nn} \end{vmatrix}$$

$$= (-1)^{n+1}a_{1n}(-1)^n a_{2,n-1}\cdots(-1)^3 a_{n-1,2}a_{n1}$$

$$= (-1)^{(n-1)(n+4)/2}a_{1n}a_{2,n-1}\cdots a_{n-1,2}a_{n1}$$

and

$$\begin{vmatrix} a_{11} & a_{12} & \cdots & a_{1,n-1} & a_{1n} \\ a_{21} & a_{22} & \cdots & a_{2,n-1} & 0 \\ \cdots & & & & \\ a_{n-1,1} & a_{n-1,2} & \cdots & 0 & 0 \\ a_{n1} & 0 & \cdots & 0 & 0 \end{vmatrix}$$

$$= (-1)^{(n-1)(n+4)/2}a_{1n}a_{2,n-1}\cdots a_{n-1,2}a_{n1}.$$

We note that for a general determinant the representation (3.5) is only of theoretical interest; in fact, it has $n!$ terms to sum and each term requires $(n-1)$ multiplications, and hence for the computation of $|A|$ we need in total $(n-1) \times n!$ multiplications, which for large n is a formidable task. To reduce the computational work considerably we often use the following fundamental properties of determinants.

1. If any row or column of A has only zero elements, then $|A| = 0$.

2. If two rows (or columns) of A are equal or have a constant ratio, then $|A| = 0$.

3. If any two consecutive rows (or columns) of A are interchanged, then the determinant of the new matrix A_1 is $-|A|$.

4. If a row (or column) of A is multiplied by a constant α, then the determinant of the new matrix A_1 is $\alpha|A|$.

5. If a constant multiple of one row (or column) of A is added to another, then the determinant of the new matrix A_1 is unchanged.

6. $|A^t| = |A|$.

7. $|AB| = |A||B| = |BA|$. This property for the determinants is very interesting. For the matrices A and B given in Example 2.2 we have seen that $AB \neq BA$. However, we have

$$|A| = \begin{vmatrix} 1 & 1 & 0 \\ -1 & 2 & 1 \\ 0 & 0 & 2 \end{vmatrix} = 6, \quad |B| = \begin{vmatrix} 2 & 3 & 4 \\ 1 & 2 & 3 \\ -1 & 1 & 2 \end{vmatrix} = -1,$$

$$|AB| = \begin{vmatrix} 3 & 5 & 7 \\ -1 & 2 & 4 \\ -2 & 2 & 4 \end{vmatrix} = -6, \quad |BA| = \begin{vmatrix} -1 & 8 & 11 \\ -1 & 5 & 8 \\ -2 & 1 & 5 \end{vmatrix} = -6.$$

Thus, $|AB| = |A||B| = |BA|$.

8. If each element of a row or a column of A is expressed as the sum of two numbers, then $|A|$ can be written as the sum of two determinants. For example,

$$\begin{vmatrix} a_{11} & a_{12}+b_{12} & a_{13} \\ a_{21} & a_{22}+b_{22} & a_{23} \\ a_{31} & a_{32}+b_{32} & a_{33} \end{vmatrix} = \begin{vmatrix} a_{11} & a_{12} & a_{13} \\ a_{21} & a_{22} & a_{23} \\ a_{31} & a_{32} & a_{33} \end{vmatrix} + \begin{vmatrix} a_{11} & b_{12} & a_{13} \\ a_{21} & b_{22} & a_{23} \\ a_{31} & b_{32} & a_{33} \end{vmatrix}.$$

9. If the elements of A are polynomial functions of x and two rows or columns become identical when $x = a$, then $x - a$ is a factor of $|A|$.

While a systematic procedure for the computation of determinants of any order will be given in Chapter 6, the following examples illustrate the usefulness of the above properties.

Example 3.3. We have

$$\Delta = \begin{vmatrix} 11 & 3 & 4 \\ 19 & 6 & 5 \\ 21 & 7 & 8 \end{vmatrix} = - \begin{vmatrix} 3 & 11 & 4 \\ 6 & 19 & 5 \\ 7 & 21 & 8 \end{vmatrix} = - \begin{vmatrix} -1 & 11 & 4 \\ 1 & 19 & 5 \\ -1 & 21 & 8 \end{vmatrix}$$

$$= \begin{vmatrix} 1 & 19 & 5 \\ -1 & 11 & 4 \\ -1 & 21 & 8 \end{vmatrix} = \begin{vmatrix} 1 & 19 & 5 \\ 0 & 30 & 9 \\ 0 & 40 & 13 \end{vmatrix} = \begin{vmatrix} 30 & 9 \\ 40 & 13 \end{vmatrix}$$

$$= 30 \times 13 - 9 \times 40 = 390 - 360 = 30.$$

Example 3.4. We have

$$\Delta = \begin{vmatrix} a-b-c & 2a & 2a \\ 2b & b-c-a & 2b \\ 2c & 2c & c-a-b \end{vmatrix} = \begin{vmatrix} a+b+c & a+b+c & a+b+c \\ 2b & b-c-a & 2b \\ 2c & 2c & c-a-b \end{vmatrix}$$

$$= (a+b+c) \begin{vmatrix} 1 & 1 & 1 \\ 2b & b-c-a & 2b \\ 2c & 2c & c-a-b \end{vmatrix}$$

$$= (a+b+c) \begin{vmatrix} 1 & 0 & 0 \\ 2b & -b-c-a & 0 \\ 2c & 0 & -c-a-b \end{vmatrix}$$

$$= (a+b+c) \begin{vmatrix} -b-c-a & 0 \\ 0 & -c-a-b \end{vmatrix} = (a+b+c)^3.$$

Example 3.5. The *Vandermonde matrix*

$$V = \begin{pmatrix} 1 & 1 & \cdots & 1 \\ x_1 & x_2 & \cdots & x_n \\ x_1^2 & x_2^2 & \cdots & x_n^2 \\ \cdots \\ x_1^{n-1} & x_2^{n-1} & \cdots & x_n^{n-1} \end{pmatrix}$$

plays an important role in polynomial interpolation theory. By induction we shall show that

$$|V| = \prod_{1 \le i < j \le n} (x_j - x_i).$$

Indeed, we have

$$\begin{vmatrix} 1 & 1 \\ x_1 & x_2 \end{vmatrix} = (x_2 - x_1)$$

and

$$|V| = \begin{vmatrix} 1 & 1 & \cdots & 1 \\ 0 & x_2 - x_1 & \cdots & x_n - x_1 \\ 0 & x_2^2 - x_2 x_1 & \cdots & x_n^2 - x_n x_1 \\ \cdots \\ 0 & x_2^{n-1} - x_2^{n-2} x_1 & \cdots & x_n^{n-1} - x_n^{n-2} x_1 \end{vmatrix}$$

$$= \prod_{j=2}^{n} (x_j - x_1) \begin{vmatrix} 1 & 1 & \cdots & 1 \\ x_2 & x_3 & \cdots & x_n \\ x_2^2 & x_3^2 & \cdots & x_n^2 \\ \cdots \\ x_2^{n-2} & x_3^{n-2} & \cdots & x_n^{n-2} \end{vmatrix}$$

$$= \prod_{j=2}^{n} (x_j - x_1) \prod_{2 \le i < j \le n} (x_j - x_i) = \prod_{1 \le i < j \le n} (x_j - x_i).$$

Example 3.6. From the definition of determinants it is clear that for a given $n \times n$ matrix $A(x) = (a_{ij}(x))$ of differentiable functions in an interval J, the function $\det A(x)$ is differentiable in J. We shall compute $(\det A(x))'$ by using the expansion of $\det A(x)$ given in (3.1). Since

$$\det A(x) = \sum_{j=1}^{n} a_{ij}(x)\alpha_{ij}(x)$$

it follows that

$$\frac{\partial \det A(x)}{\partial a_{ij}(x)} = \alpha_{ij}(x)$$

and hence

$$(\det A(x))' = \sum_{j=1}^{n}\sum_{i=1}^{n} \frac{\partial \det A(x)}{\partial a_{ij}(x)} \frac{da_{ij}(x)}{dx} = \sum_{j=1}^{n}\sum_{i=1}^{n} \alpha_{ij}(x)a'_{ij}(x),$$

which is equivalent to

$$(\det A(x))' = \begin{vmatrix} a'_{11}(x) & \cdots & a'_{1n}(x) \\ a_{21}(x) & \cdots & a_{2n}(x) \\ & \cdots & \\ a_{n1}(x) & \cdots & a_{nn}(x) \end{vmatrix} + \begin{vmatrix} a_{11}(x) & \cdots & a_{1n}(x) \\ a'_{21}(x) & \cdots & a'_{2n}(x) \\ & \cdots & \\ a_{n1}(x) & \cdots & a_{nn}(x) \end{vmatrix} + \cdots$$

$$+ \begin{vmatrix} a_{11}(x) & \cdots & a_{1n}(x) \\ a_{21}(x) & \cdots & a_{2n}(x) \\ & \cdots & \\ a'_{n1}(x) & \cdots & a'_{nn}(x) \end{vmatrix}. \tag{3.6}$$

Problems

3.1. Evaluate

(i) $\begin{vmatrix} 21 & 17 & 7 & 10 \\ 24 & 22 & 6 & 10 \\ 6 & 8 & 2 & 3 \\ 5 & 7 & 1 & 2 \end{vmatrix}$,   (ii) $\begin{vmatrix} 1+i & 1-i & i \\ 1-i & i & 1+i \\ i & 1+i & 1-i \end{vmatrix}$.

3.2. Solve the equations

(i) $\begin{vmatrix} x+2 & 2x+3 & 3x+4 \\ 2x+3 & 3x+4 & 4x+5 \\ 3x+5 & 5x+8 & 10x+17 \end{vmatrix} = 0$,   (ii) $\begin{vmatrix} 1+x & 2 & 3 \\ 1 & 2+x & 3 \\ 1 & 2 & 3+x \end{vmatrix} = 0$.

3.3. Show that

(i) $\begin{vmatrix} 1+a & 1 & 1 & 1 \\ 1 & 1+b & 1 & 1 \\ 1 & 1 & 1+c & 1 \\ 1 & 1 & 1 & 1+d \end{vmatrix} = abcd\left(1 + \frac{1}{a} + \frac{1}{b} + \frac{1}{c} + \frac{1}{d}\right)$

(ii) $\begin{vmatrix} a^2+\lambda & ab & ac & ad \\ ab & b^2+\lambda & bc & bd \\ ac & bc & c^2+\lambda & cd \\ ad & bd & cd & d^2+\lambda \end{vmatrix} = \lambda^3(a^2+b^2+c^2+d^2+\lambda)$

(iii) $\begin{vmatrix} 1 & 1 & 1 & \cdots & 1 \\ 1 & 2 & 2^2 & \cdots & 2^{n-1} \\ 1 & 3 & 3^2 & \cdots & 3^{n-1} \\ \cdots & & & & \\ 1 & n & n^2 & \cdots & n^{n-1} \end{vmatrix} = 1!2!3!\cdots(n-1)!.$

3.4. Show that

(i) if the matrix $A \in M^{n\times n}$ is skew-symmetric, then $\det(A) = (-1)^n \times \det(A)$, and hence $\det(A) = 0$ if n is odd

(ii) if the matrix $A \in M^{n\times n}$ is hermitian, then $\det(A)$ is real.

3.5. Let A and B be $n \times n$ matrices such that $AB = BA$. Show that $\det(A^2 + B^2) \geq 0$.

Answers or Hints

3.1. (i) 0.
(ii) $4 + 7i$.

3.2. (i) $-1, -1, -2$.
(ii) $0, 0, -6$.

3.3. (i) $\begin{vmatrix} 1+a & -a & -a & -a \\ 1 & b & 0 & 0 \\ 1 & 0 & c & 0 \\ 1 & 0 & 0 & d \end{vmatrix} = \begin{vmatrix} 1+a+\frac{a}{b}+\frac{a}{c}+\frac{a}{d} & -a & -a & -a \\ 0 & b & 0 & 0 \\ 0 & 0 & c & 0 \\ 0 & 0 & 0 & d \end{vmatrix}$

(ii) $a^2b^2c^2d^2 \begin{vmatrix} \frac{a^2+\lambda}{a^2} & 1 & 1 & 1 \\ 1 & \frac{b^2+\lambda}{b^2} & 1 & 1 \\ 1 & 1 & \frac{c^2+\lambda}{c^2} & 1 \\ 1 & 1 & 1 & \frac{d^2+\lambda}{d^2} \end{vmatrix}$

(iii) See Example 3.5.

3.4. (i) $\det(A) = \det(-A^t) = (-1)^n\det(A^t) = (-1)^n\det(A)$

(ii) $\det(A) = \det(\overline{A}^t) = \det(\overline{A}) = \sum_{\sigma\in S_n}(\text{sgn }\sigma)\overline{a}_{1\sigma(1)}\overline{a}_{2\sigma(2)}\cdots\overline{a}_{n\sigma(n)}$
$= \overline{\sum_{\sigma\in S_n}(\text{sgn }\sigma)a_{1\sigma(1)}a_{2\sigma(2)}\cdots a_{n\sigma(n)}} = \overline{\det(A)}.$

3.5. $\det(A^2+B^2) = \det(A+iB)(A-iB) = \det(A+iB)\det(A-iB) = \det(A+iB)\overline{\det(A+iB)} \geq 0.$

Chapter 4

Invertible Matrices

In this chapter we shall show that the theory of determinants can be applied to find the inverse of a given square matrix. In particular, we shall provide analytical representations of inverses of some band matrices that are of immense value in chemistry, physics, and solving two-point boundary value problems for ordinary differential equations by finite difference methods.

A square matrix $A = (a_{ij}) \in M^{n \times n}$ is said to be *invertible* or *nonsingular* if and only if there exists a matrix $B \in M^{n \times n}$ such that $AB = BA = I$. Such a matrix B is unique. Indeed, if $AB = BA = I$ and $AC = CA = I$, then $B = BI = B(AC) = (BA)C = IC = C$. The matrix B is called the *inverse* of A and is denoted by A^{-1}. If A is nonsingular, then since $AA^{-1} = A^{-1}A = I$ it follows that $(A^{-1})^{-1} = A$. It is clear that if B is the inverse of A, then A is the inverse of B. A square matrix is called *singular* if it has no inverse. We begin with the following result whose proof follows from the discussion in our next chapter (see Problem 5.1).

Theorem 4.1. If matrices $A = (a_{ij})$ and $B = (b_{ij})$ are in $M^{n \times n}$ such that $AB = I$ or $BA = I$, then A and B both are invertible, and each is the inverse of the other.

Thus to find the inverse of A it suffices to find the matrix B such that $AB = I$ or $BA = I$, i.e., we do not have to check both equalities.

Example 4.1. We shall find the inverse

$$B = \begin{pmatrix} x_1 & x_2 \\ y_1 & y_2 \end{pmatrix}$$

of the matrix

$$A = \begin{pmatrix} a & b \\ c & d \end{pmatrix}.$$

For this, we note that $AB = I$ is the same as the systems

$$\begin{array}{ll} ax_1 + by_1 = 1, & ax_2 + by_2 = 0 \\ cx_1 + dy_1 = 0, & cx_2 + dy_2 = 1. \end{array}$$

If $|A| = ad - bc \neq 0$, then these systems can be solved uniquely, to obtain

$$x_1 = \frac{d}{|A|}, \quad y_1 = -\frac{c}{|A|}, \quad x_2 = -\frac{b}{|A|}, \quad y_2 = \frac{a}{|A|}.$$

Thus, the inverse B of the matrix A exists if and only if $|A| \neq 0$, and it appears as

$$B = \begin{pmatrix} d/|A| & -b/|A| \\ -c/|A| & a/|A| \end{pmatrix} = \frac{1}{|A|} \begin{pmatrix} d & -b \\ -c & a \end{pmatrix}.$$

From Example 4.1 it is clear that for a given matrix the inverse may not exist, and hence there are singular as well as nonsingular matrices.

Theorem 4.2. If $A = (a_{ij})$ and $B = (b_{ij})$ are nonsingular matrices in $M^{n \times n}$, then AB is also nonsingular, and $(AB)^{-1} = B^{-1}A^{-1}$. Conversely, if AB is nonsingular, then A and B are nonsingular.

Proof. Since A^{-1} and B^{-1} exist,

$$(AB)(B^{-1}A^{-1}) = A(BB^{-1})A^{-1} = A(I)A^{-1} = AA^{-1} = I.$$

Thus Theorem 4.1 implies that AB is nonsingular, and $(AB)^{-1} = B^{-1}A^{-1}$. Conversely, in view of Theorem 4.1 it suffices to note that

$$I = (AB)(AB)^{-1} = A(B(AB)^{-1})$$

and

$$I = (AB)^{-1}(AB) = ((AB)^{-1}A)B. \quad \blacksquare$$

Corollary 4.1. If A is an invertible matrix in $M^{n \times n}$, then

$$\det(A^{-1}) = (\det(A))^{-1}.$$

Proof. Since $AA^{-1} = I$, $\det(AA^{-1}) = \det(A)\det(A^{-1}) = 1$.

Corollary 4.2. If $A, B, C \in M^{n \times n}$ and A is an invertible matrix, then $AB = AC$ implies $B = C$.

Corollary 4.3. If $A_i \in M^{n \times n}$, $i = 1, 2, \cdots, m$ are nonsingular matrices, then $A_1 A_2 \cdots A_m$ is nonsingular, and

$$(A_1 A_2 \cdots A_m)^{-1} = A_m^{-1} A_{m-1}^{-1} \cdots A_1^{-1}.$$

Theorem 4.3. If A is a nonsingular real (complex) matrix, then $A^t (A^H)$ are nonsingular, and $(A^t)^{-1} = (A^{-1})^t ((A^H)^{-1} = (A^{-1})^H)$.

Proof. Since $AA^{-1} = I$, we have $(A^{-1})^t A^t = I$, and similarly from $A^{-1}A = I$ it follows that $(A^t)(A^{-1})^t = I$. These relations imply that $(A^{-1})^t = (A^t)^{-1}$. $\quad \blacksquare$

Theorem 4.4. The matrix $A \in M^{n \times n}$ is nonsingular if and only if $\det(A) \neq 0$.

The *adjoint* of a given matrix $A \in M^{n \times n}$, written as adj A, is the transpose of the matrix of cofactors of A, i.e., adj $A = (\alpha_{ij})^t = (\alpha_{ji})$.

Theorem 4.5. If $A \in M^{n \times n}$ is a nonsingular matrix, then

$$A^{-1} = \frac{\text{adj } A}{|A|}. \tag{4.1}$$

Proof. From (3.1)–(3.3), we have $A \, (\text{adj } A) = (c_{ij})$, where

$$c_{ij} = \sum_{k=1}^{n} a_{ik} \alpha_{jk} = \begin{cases} \det(A) & \text{if } i = j \\ 0 & \text{if } i \neq j. \end{cases}$$

Hence, $A \, (\text{adj } A) = \det(A) I$, which implies (4.1). ∎

Example 4.2. For the matrix

$$A = \begin{pmatrix} 2 & -1 & 1 \\ 3 & 2 & -5 \\ 1 & 3 & -2 \end{pmatrix},$$

we have

$$|A| = \begin{vmatrix} 2 & -1 & 1 \\ 3 & 2 & -5 \\ 1 & 3 & -2 \end{vmatrix} = 28 \quad \text{and} \quad \text{adj } A = \begin{pmatrix} 11 & 1 & 3 \\ 1 & -5 & 13 \\ 7 & -7 & 7 \end{pmatrix}$$

and hence from (4.1) it follows that

$$A^{-1} = \begin{pmatrix} \frac{11}{28} & \frac{1}{28} & \frac{3}{28} \\ \frac{1}{28} & -\frac{5}{28} & \frac{13}{28} \\ \frac{1}{4} & -\frac{1}{4} & \frac{1}{4} \end{pmatrix}.$$

An $n \times n$ matrix $A = (a_{ij})$ is said to be a *band matrix* if there exist integers r and s, $1 < r, s < n$ such that $a_{ij} = 0$ for all $j - i \geq r$ and $i - j \geq s$. The number $w = r + s - 1$ is called the bandwidth of A. Matrices with $r = s = 2$ so that $w = 3$ are called *tridiagonal* matrices. We shall find the inverse of the tridiagonal matrix

$$A_n(x) = \begin{pmatrix} x & -1 & & & & \\ -1 & x & -1 & & & \\ & -1 & x & -1 & & \\ & & \cdots & \cdots & & \\ & & & -1 & x & -1 \\ & & & & -1 & x \end{pmatrix}, \tag{4.2}$$

where $x \geq 0$. For this, let $D_n(x) = D_n = |A_n(x)|$. Then, D_n satisfies the following second order linear difference equation

$$D_n = x D_{n-1} - D_{n-2}, \quad n = 1, 2, \cdots \tag{4.3}$$

together with the initial conditions

$$D_{-1} = 0, \quad D_0 = 1. \tag{4.4}$$

The general solution of (4.3) can be written as

$$D_n = A\lambda_1^n + B\lambda_2^n, \tag{4.5}$$

where λ_1, λ_2 are the roots of the equation $\lambda^2 - x\lambda + 1 = 0$. Using (4.4), it follows that

$$D_n = \begin{cases} \sinh(n+1)\theta/\sinh\theta & \text{if} \quad x = 2\cosh\theta > 2 \\ n+1 & \text{if} \quad x = 2 \\ \sin(n+1)\theta/\sin\theta & \text{if} \quad 0 \le x = 2\cos\theta < 2. \end{cases} \tag{4.6}$$

Now let $B_n = (b_{ij})$ be the inverse of $A_n(x)$, i.e., $A_n(x)B_n = I$. Then, for $1 \le j \le n$ it follows that

$$-b_{i-1,j} + x b_{ij} - b_{i+1,j} = \begin{cases} 0, & i = 1, \cdots, j-1, \ b_{0j} = 0 \\ 1, & i = j \\ 0, & i = j+1, \cdots, n, \ b_{n+1,j} = 0. \end{cases} \tag{4.7}$$

Let $x = 2\cosh\theta > 2$. When $1 \le i \le j - 1$, the general solution of (4.7) is $b_{ij} = A\lambda_1^i + B\lambda_2^i$, where once again λ_1, λ_2 are the roots of the equation $\lambda^2 - x\lambda + 1 = 0$. This solution is valid for $0 \le i \le j$, and in view of $b_{0j} = 0$ and (4.6) can be written as

$$b_{ij} = 2A\sinh i\theta = 2AD_{i-1}\sinh\theta, \quad 0 \le i \le j. \tag{4.8}$$

Similarly, when $j + 1 \le i \le n$, the general solution of (4.7) is $b_{ij} = C\lambda_1^i + D\lambda_2^i$, which is valid for $j \le i \le n + 1$. This solution in view of $b_{n+1,j} = 0$ and (4.6) appears as

$$b_{ij} = -2CD_{n-i}e^{(n+1)\theta}\sinh\theta, \quad j \le i \le n + 1. \tag{4.9}$$

Now equating (4.8) and (4.9) for $i = j$, we obtain the equation

$$AD_{j-1} + Ce^{(n+1)\theta}D_{n-j} = 0. \tag{4.10}$$

Also, substituting (4.8) and (4.9) in (4.7) with $i = j$, we get the equation

$$\sinh\theta \left[AD_j + Ce^{(n+1)\theta}D_{n-j-1} \right] = \frac{1}{2}. \tag{4.11}$$

Solving (4.10) and (4.11), we find

$$A = \frac{D_{n-j}}{2D_n\sinh\theta}, \quad C = -\frac{D_{j-1}}{2D_ne^{(n+1)\theta}\sinh\theta}. \tag{4.12}$$

Now substituting (4.12) in (4.8) and (4.9), we obtain

$$b_{ij} = \frac{1}{D_n} \begin{cases} D_{i-1}D_{n-j}, & i \le j \\ D_{j-1}D_{n-i}, & i \ge j. \end{cases} \tag{4.13}$$

Finally, we remark that the above calculations can be modified to show that the formula (4.13) holds for $0 \le x \le 2$ also.

We summarize the above result in the following theorem.

Theorem 4.6. For the tridiagonal matrix $A_n(x)$, $x \ge 0$ given in (4.2), the inverse matrix $A_n^{-1}(x) = B_n = (b_{ij})$ is symmetric, i.e., $b_{ij} = b_{ji}$ and

$$b_{ij} = \frac{D_{i-1}D_{n-j}}{D_n} > 0, \quad i \le j. \tag{4.14}$$

Example 4.3. To compute $A_5^{-1}(3)$, first we use (4.3), (4.4), to obtain

$$D_1 = 3, \quad D_2 = 8, \quad D_3 = 21, \quad D_4 = 55, \quad D_5 = 144$$

and then use (4.14) to get

$$A_5^{-1}(3) = B_5 = \frac{1}{144} \begin{pmatrix} 55 & 21 & 8 & 3 & 1 \\ 21 & 63 & 24 & 9 & 3 \\ 8 & 24 & 64 & 24 & 8 \\ 3 & 9 & 24 & 63 & 21 \\ 1 & 3 & 8 & 21 & 55 \end{pmatrix}.$$

Now let $x = -y$, $y \ge 0$. Then, we have

$$A_n(x) = A_n(-y) = -\mathcal{A}_n(y), \tag{4.15}$$

where

$$\mathcal{A}_n(y) = \begin{pmatrix} y & 1 & & & & \\ 1 & y & 1 & & & \\ & 1 & y & 1 & & \\ & & \cdots & \cdots & & \\ & & & & 1 & y & 1 \\ & & & & & 1 & y \end{pmatrix}, \tag{4.16}$$

For the matrix (4.16) a result analogous to Theorem 4.6 is the following:

Theorem 4.7. For the tridiagonal matrix $\mathcal{A}_n(y)$, $y \ge 0$ given in (4.16), the inverse matrix $\mathcal{A}_n^{-1}(y) = \mathcal{B}_n = (\beta_{ij})$ is symmetric, i.e., $\beta_{ij} = \beta_{ji}$ and

$$\beta_{ij} = \frac{(-1)^{i+j}D_{i-1}D_{n-j}}{D_n}, \quad i \le j. \tag{4.17}$$

Problems

4.1. Which of the following matrices are singular?

$$A = \begin{pmatrix} 1 & 2 & 3 \\ 1 & 1 & 2 \\ 1 & 3 & 4 \end{pmatrix}, \quad B = \begin{pmatrix} 1 & 1 & 1 \\ 2 & 4 & 8 \\ 3 & 9 & 25 \end{pmatrix}, \quad C = \begin{pmatrix} 2 & 5 & 19 \\ 1 & -2 & -4 \\ -3 & 2 & 0 \end{pmatrix}$$

Use (4.1) to find the inverses of nonsingular matrices.

4.2. For what values of x is the following matrix singular?

$$A = \begin{pmatrix} 3-x & 2 & 2 \\ 2 & 4-x & 1 \\ -2 & -4 & -1-x \end{pmatrix}$$

4.3. Let

$$A = \begin{pmatrix} -\dfrac{1}{2} & -\dfrac{\sqrt{3}}{2} & 0 \\ -\dfrac{\sqrt{3}}{2} & \dfrac{1}{2} & 0 \\ 0 & 0 & 0 \end{pmatrix}, \quad P = \begin{pmatrix} \dfrac{1}{2} & \dfrac{\sqrt{3}}{2} & 0 \\ -\dfrac{\sqrt{3}}{2} & \dfrac{1}{2} & 0 \\ 0 & 0 & 1 \end{pmatrix}.$$

Show that $P^{-1}AP$ is a diagonal matrix.

4.4. Show that the inverse of an upper (lower) triangular nonsingular square matrix is an upper (lower) triangular square matrix. In particular, show that the inverse of a diagonal matrix A with nonzero diagonal elements a_{ii}, $i = 1, \cdots, n$ is a diagonal matrix A^{-1} with diagonal elements $1/a_{ii}$, $i = 1, \cdots, n$.

4.5. Let the square matrices A, B and $A + B$ be nonsingular. Show that $A^{-1} + B^{-1}$ is nonsingular, and

$$(A^{-1} + B^{-1})^{-1} = A(A+B)^{-1}B = B(A+B)^{-1}A.$$

4.6. Let $A, B \in C^{n \times n}$. Show that if B is nonsingular, then $\text{tr}(B^{-1}AB) = \text{tr}(A)$.

4.7. A real square matrix A is called *orthogonal* if and only if $A^t = A^{-1}$, i.e., $AA^t = A^tA = I$. Thus, $\det(A) = \pm 1$. A complex square matrix A is said to be *unitary* if and only if $A^H = A^{-1}$, i.e., $A^H A^{-1} = A^{-1}A^H = I$. If $A, B \in M^{n \times n}$ are unitary matrices, show that A^H, A^{-1}, AB are also unitary matrices.

4.8. Verify that the following matrices are orthogonal:

$$\text{(i)} \begin{pmatrix} -\dfrac{2}{3} & \dfrac{1}{3} & \dfrac{2}{3} \\ \dfrac{2}{3} & \dfrac{2}{3} & \dfrac{1}{3} \\ \dfrac{1}{3} & -\dfrac{2}{3} & \dfrac{2}{3} \end{pmatrix}, \quad \text{(ii)} \begin{pmatrix} \dfrac{1}{2} & \dfrac{1}{2} & -\dfrac{1}{2} & -\dfrac{1}{2} \\ \dfrac{1}{3\sqrt2} & \dfrac{2}{3} & \dfrac{2}{3} & \dfrac{1}{3\sqrt2} \\ \dfrac{1}{2} & -\dfrac{1}{2} & \dfrac{1}{2} & -\dfrac{1}{2} \\ \dfrac{2}{3} & -\dfrac{1}{3\sqrt2} & -\dfrac{1}{3\sqrt2} & \dfrac{2}{3} \end{pmatrix}.$$

4.9. Let A be a skew-symmetric matrix and $I - A$ is nonsingular. Show that the matrix $B = (I + A)(I - A)^{-1}$ is orthogonal.

4.10. Let W be a $1 \times n$ matrix (row vector) such that $WW^t = 1$. The $n \times n$ matrix $H = I - 2W^t W$ is called a *Householder matrix*. Show that H is symmetric and orthogonal.

4.11. Let A, B be real square matrices, and let the matrix P be orthogonal and $B = P^{-1}AP$. Show that $\text{tr}(A^t A) = \text{tr}(B^t B)$.

4.12. For the matrices $A_n(x)$, $x \geq 0$ and $\mathcal{A}_n(y)$, $y \geq 0$ given in (4.2) and (4.16), show that

(i) $A_n(x)\mathcal{A}_n(y) = \mathcal{A}_n(y)A_n(x)$

(ii) $[A_n(x)\mathcal{A}_n(y)]^{-1} = \dfrac{1}{x+y}\left[A_n^{-1}(x) + \mathcal{A}_n^{-1}(y)\right], \; x+y \neq 0$

(iii) $[A_n(x)\mathcal{A}_n(y)]^{-1} = \dfrac{1}{x-y}\left[A_n^{-1}(y) - A_n^{-1}(x)\right], \; x-y \neq 0$

(iv) $[\mathcal{A}_n(x)\mathcal{A}_n(y)]^{-1} = \dfrac{1}{x-y}[\mathcal{A}_n^{-1}(y) - \mathcal{A}_n^{-1}(x)], \; x-y \neq 0.$

(v) the matrix $[A_n^2(x)]^{-1} = C = (c_{ij})$ is symmetric, and

$$\begin{aligned} c_{ij} &= \frac{2}{(x^2-4)D_n^2}\left[D_n D_{n-j}\left(\frac{i+1}{2}D_{i-2} - \frac{i-1}{2}D_i\right) \right. \\ &\quad \left. + D_{i-1}\left(\frac{j}{2}D_{2n+1-j} - \frac{2n+2-j}{2}D_{j-1}\right)\right], \quad x \neq 2, \; i \leq j \\ &= \frac{i(n-j+1)}{6(n+1)}[2j(n+1) - (i^2+j^2) + 1], \quad x = 2, \; i \leq j \end{aligned}$$

(vi) the matrix $[\mathcal{A}_n^2(y)]^{-1} = \Gamma = (\gamma_{ij})$ is symmetric, and $\gamma_{ij} = (-1)^{i+j}c_{ij}$.

4.13. Consider the tridiagonal matrix of order n,

$$A_n(x, y) = \begin{pmatrix} 1+x & -1 & & & & & \\ -1 & 2 & -1 & & & & \\ & -1 & 2 & -1 & & & \\ & & \ddots & \ddots & & & \\ & & & & -1 & 2 & -1 \\ & & & & & -1 & 1+y \end{pmatrix},$$

where $x + y + (n-1)xy \neq 0$. Show that $A_n^{-1}(x, y) = (b_{ij})$ is symmetric and

$$b_{ij} = \frac{[1 + (i-1)x][1 + (n-j)y]}{x + y + (n-1)xy}, \quad i \leq j.$$

4.14. Consider the tridiagonal matrix of order n,

$$A_n(x, y) = \begin{pmatrix} 1 & -y & & & & & \\ -x & 1 & -y & & & & \\ & -x & 1 & -y & & & \\ & & \ddots & \ddots & & & \\ & & & & -x & 1 & -y \\ & & & & & -x & 1 \end{pmatrix},$$

where $x \geq 0$, $y \geq 0$. Show that

(i)   its determinant D_n is

$$D_n = \begin{cases} 1, & xy = 0 \\ (xy)^{n/2}\dfrac{\sinh(n+1)\theta}{\sinh\theta}, & \cosh\theta = \dfrac{1}{2\sqrt{xy}}, \quad 0 < xy < \dfrac{1}{4} \\ \dfrac{n+1}{2^n}, & xy = \dfrac{1}{4} \\ (xy)^{n/2}\dfrac{\sin(n+1)\theta}{\sin\theta}, & \cos\theta = \dfrac{1}{2\sqrt{xy}}, \quad xy > \dfrac{1}{4} \end{cases}$$

(ii)   the elements b_{ij} of its inverse matrix B are

$$b_{ij} = \frac{1}{D_n} \begin{cases} y^{j-i}D_{i-1}D_{n-j}, & i \leq j \\ x^{i-j}D_{j-1}D_{n-i}, & i \geq j. \end{cases}$$

4.15. Consider the matrix $\mathcal{A}_n(x, y) = A_n(-x, -y)$, where $A_n(x, y)$ is the same as in Problem 4.14. Show that the elements α_{ij} of the inverse matrix $\mathcal{A}_n^{-1}(x, y)$ are

$$\alpha_{ij} = \frac{1}{D_n} \begin{cases} (-1)^{i+j}y^{j-i}D_{i-1}D_{n-j}, & i \leq j \\ (-1)^{i+j}x^{i-j}D_{j-1}D_{n-i}, & i \geq j \end{cases}$$

where D_n is the same as in Problem 4.14.

4.16. The $n \times n$ matrix $A_n = (a_{ij})$, where

$$a_{ij} = \begin{cases} 1, & j - i = 1 \\ 1, & i = j \\ -1, & i - j = 1 \\ 0, & \text{otherwise} \end{cases}$$

is called the *Fibonacci matrix*. Let $F_n = \det(A_n)$ to show that the *Fibonacci numbers* F_n satisfy

(i) $F_n = F_{n-1} + F_{n-2}, \ n = 1, 2, \cdots$ where $F_{-1} = 0, \ F_0 = 1$

(ii) $F_n = \dfrac{1}{2^{n+1}\sqrt{5}} \left[(1 + \sqrt{5})^{n+1} - (1 - \sqrt{5})^{n+1} \right].$

Also, find the inverse of the matrix A_n.

4.17. An $n \times n$ matrix is called *nilpotent* if $A^k = 0$ for some positive integer k. Show that

(i)   the following matrix is nilpotent

$$A = \begin{pmatrix} 2 & 11 & 3 \\ -2 & -11 & -3 \\ 8 & 35 & 9 \end{pmatrix}$$

(ii) every nilpotent matrix is singular

(iii) if A is nilpotent, then $I - A$ is nonsingular

(iv) if the matrices A, B are nilpotent and $AB = BA$, then AB and $A + B$ are nilpotent.

4.18. An $n \times n$ matrix is called *idempotent* if $A^2 = A$. Show that

(i)   matrices I and 0 are idempotent

(ii) if A is idempotent, then A^t and $I - A$ are idempotent

(iii) every idempotent matrix except I is singular

(iv) if A is idempotent, then $2A - I$ is invertible and is its own inverse

(v)   if the matrices A, B are idempotent and $AB = BA$, then AB is idempotent.

4.19. An $n \times n$ matrix that results from permuting the rows of an $n \times n$ identity matrix is called a *permutation matrix*. Thus, each permutation matrix has 1 in each row and each column and all other elements are 0. Show that

(i) each permutation matrix P is nonsingular and orthogonal

(ii) product of two permutation matrices is a permutation matrix.

4.20. Show that the invertible $n \times n$ matrices do not form a subspace of $M^{n \times n}$.

Answers or Hints

4.1. A is singular. B is nonsingular, and $B^{-1} = \dfrac{1}{8} \begin{pmatrix} 28 & -16 & 4 \\ -26 & 22 & -6 \\ 6 & -6 & 2 \end{pmatrix}$. C is singular.

4.2. $\det A = 6x^2 - 9x - x^3$ so $x = 0, 3, 3$.

4.3. $P^{-1} = \begin{pmatrix} \frac{1}{2} & -\frac{\sqrt{3}}{2} & 0 \\ \frac{\sqrt{3}}{2} & \frac{1}{2} & 0 \\ 0 & 0 & 1 \end{pmatrix}$ so $P^{-1}AP = \begin{pmatrix} 1 & 0 & 0 \\ 0 & -1 & 0 \\ 0 & 0 & 0 \end{pmatrix}$.

4.4. If A is upper triangular, then $a_{ij} = 0$, $i > j$, and hence $\alpha_{ij} = 0$, $j > i$. Now use (4.1).

4.5. $(B(A+B)^{-1}A)^{-1} = A^{-1}(A+B)B^{-1} = A^{-1}(AB^{-1}+I) = B^{-1} + A^{-1}$.

4.6. Use $\operatorname{tr}(AB) = \operatorname{tr}(BA)$ to get $\operatorname{tr}(B^{-1}AB) = \operatorname{tr}(ABB^{-1})$.

4.7. $(A^H)^H = (A^{-1})^H = (A^H)^{-1}$. Now A^{-1} is unitary, as follows from $A^H = A^{-1}$. Finally, since $A^H = A^{-1}$ and $B^H = B^{-1}$, we have $(AB)(AB)^H = ABB^H A^H = ABB^{-1}A^{-1} = I$. Thus, $(AB)^H = (AB)^{-1}$, and so AB is unitary.

4.8. (i) For the given matrix the inverse is $\begin{pmatrix} -\frac{2}{3} & \frac{2}{3} & \frac{1}{3} \\ \frac{1}{3} & -\frac{1}{3} & -\frac{2}{3} \\ \frac{2}{3} & \frac{1}{3} & \frac{2}{3} \end{pmatrix}$.

(ii)   For the given matrix the inverse is $\begin{pmatrix} \frac{1}{2} & \frac{1}{3\sqrt{2}} & \frac{1}{2} & \frac{2}{3} \\ \frac{1}{2} & -\frac{2}{3} & -\frac{1}{2} & -\frac{1}{3\sqrt{2}} \\ -\frac{1}{2} & \frac{2}{3} & \frac{1}{2} & -\frac{1}{3\sqrt{2}} \\ -\frac{1}{2} & \frac{1}{3\sqrt{2}} & -\frac{1}{2} & \frac{2}{3} \end{pmatrix}$.

4.9. Since $(I + A)(I - A) = (I - A)(I + A)$, we have $(I - A) = (I + A)^{-1}(I - A)(I + A)$, therefore $(I - A)(I + A)^{-1} = (I + A)^{-1}(I - A)$. Clearly, $(I - A)^t = (I - A^t)$ and $(I + A)^t = (I + A^t)$, thus $B^t = ((I + A)(I - A)^{-1})^t = ((I - A)^t)^{-1}(I + A)^t = (I - A^t)^{-1}(I + A^t) = (I + A)^{-1}(I - A) = B^{-1}$.

4.10 (i) $H^t = (I - 2W^tW)^t = I - 2(W^t)(W^t)^t = I - 2W^tW = H$.

(ii) $H^t H = H^2 = (I - 2W^tW)(I - W^tW) = I$.

4.11. We use $\operatorname{tr}(AB) = \operatorname{tr}(BA)$, to have $\operatorname{tr}(B^tB) = \operatorname{tr}((P^{-1}AP)^t(P^{-1}AP)) = \operatorname{tr}(P^tA^t(P^{-1})^tP^{-1}AP) = \cdots = \operatorname{tr}(A^tP^t(P^{-1})^tAP^{-1}P) = \operatorname{tr}(A^tA)$.

4.12. (i) Follows directly by computing both the sides.

(ii)   Using (4.15), we have $A_n(x)A_n(y)[A_n^{-1}(x) + A_n^{-1}(y)] = A_n(x)A_n(y) \times A_n^{-1}(x) + A_n(x)A_n(y)A_n^{-1}(y) = -A_n(x)A_n(-y)A_n^{-1}(x) + A_n(x) = -A_n(-y) \times A_n(x)A_n^{-1}(x) + A_n(x) = A_n(x) - A_n(-y) = (x + y)I$.

(iii)   Follow (ii).

(iv)   Follow (ii).

(v)   Use (iii), (4.14) and L'Hôpital's rule.

(vi)   Use (iv), (4.17) and L'Hôpital's rule.

4.13. Similar to that of (4.2).

4.14. Similar to that of (4.2).

4.15. Similar to that of (4.2).

4.16. Similar to that of (4.2).

4.17. (i) $A^3 = 0$.

(ii)   Since $A^k = 0$, $0 = \det(A^k) = (\det(A))^k$. Hence, $\det(A) = 0$.

(iii) $-I = A^k - I = (A - I)(A^{k-1} + A^{k-2} + \cdots + I)$, thus $(A - I)^{-1} = -(A^{k-1} + A^{k-2} + \cdots + I)$.

(iv)   If $A^k = B^\ell = 0$, for $r = \max\{k, \ell\}$, we have $(AB)^r = A^r B^r = 0$ and $(A+B)^{k+\ell} = A^{k+\ell} + \binom{k+\ell}{1}A^{k+\ell-1}B + \binom{k+\ell}{2}A^{k+\ell-2}B^2 + \cdots + \binom{k+\ell}{k+\ell}B^{k+\ell} = 0$.

4.18. (i) Clear from the definition.

(ii)   Since $A^2 = A$, $A^t = (A^2)^t = (AA)^t = A^t A^t = (A^t)^2$, $(I - A)^2 = I - A - A + A^2 = I - A$.

(iii)   If $A \neq I$, then for some $u \in R^n$, $v = Au$ where $v \neq u$. But, then $Av = A^2u = Au$, i.e., $A(v-u) = 0$. Now if A^{-1} exists, then $A^{-1}A(v-u) = v-u = 0$.

(iv) $(2A - I)(2A - I) = 4A^2 - 2A - 2A + I = I$.

(v) $(AB)^2 = (BA)(AB) = BA^2B = (BA)B = (AB)B = AB^2 = AB$.

4.19. (i) Let P be a permutation matrix. From the definition of a permutation matrix and the property 3 of determinants (Chapter 3), it is clear that $\det(P)$ is either 1 or -1. Thus, P as well as P^t both are nonsingular. Clearly, there are $n!$ permutation matrices of order $n \times n$. Now note that every permutation matrix is symmetric, i.e., $P^t = P$. We also note that interchanging two rows is a self-reverse operation, and hence every permutation matrix agrees with its inverse, i.e., $P = P^{-1}$, or $P^2 = I$. Hence, we have $P^t = P^{-1}$, which means the matrix P is orthogonal.

(ii)   If σ and π are two permutations of $N = \{1, 2, \cdots, n\}$ and P_σ and P_π are the corresponding permutation matrices, then from matrix multiplication it follows that $P_\sigma P_\pi = P_{\sigma \circ \pi}$, which implies that $P_\sigma P_\pi$ is a permutation matrix.

4.20. Consider the invertible matrices $A = \begin{pmatrix} 1 & 3 \\ 3 & 5 \end{pmatrix}$ and $B = \begin{pmatrix} -1 & 3 \\ -3 & 5 \end{pmatrix}$.

Chapter 5

Linear Systems

Systems of linear algebraic equations arise in many diverse disciplines, such as biology, business, engineering, social sciences, and statistics. Therefore, understanding the basic theory and finding efficient methods for the solutions of such systems is of great importance. We shall devote this and the next chapters to study linear systems.

Let F be a given infinite field. Consider the *nonhomogeneous* linear system of m equations in n unknowns

$$
\begin{array}{rcl}
a_{11}x_1 + a_{12}x_2 + \cdots + a_{1n}x_n & = & b_1 \\
a_{21}x_1 + a_{22}x_2 + \cdots + a_{2n}x_n & = & b_2 \\
\vdots & & \\
a_{m1}x_1 + a_{m2}x_2 + \cdots + a_{mn}x_n & = & b_m,
\end{array}
\tag{5.1}
$$

where a_{ij}, $b_i \in F$. In matrix form this system can be written as

$$
Ax = b,
\tag{5.2}
$$

where the matrix $A = (a_{ij})_{m \times n}$ and the vectors $x = (x_1, x_2, \cdots, x_n)^t$, and $b = (b_1, b_2, \cdots, b_m)^t$. By a solution x to (5.1) we mean an n-tuple that satisfies (5.1). The system (5.1) is said to be *consistent* if it has a solution, otherwise it is called *inconsistent*. If $b = 0$, (5.2) reduces to the *homogeneous* system

$$
Ax = 0.
\tag{5.3}
$$

For the homogeneous system (5.3) the zero vector $0 = (0, 0, \cdots, 0)^t$ is always a solution. This solution is called the *trivial solution* of (5.3). Clearly, the system (5.3) besides the trivial solution may also have other solutions. Such solutions we call *nontrivial solutions* of (5.3). Let F_n be the set of all solutions of (5.3), i.e., $F_n = \{x : x \in F^n \text{ such that } Ax = 0\}$, then it follows that (F_n, F) with the same addition and scalar multiplication as in Example 1.2 is a vector space. We call (F_n, F) the *solution space* of the homogeneous system (5.3), or the *null space*, or the *kernel* of the matrix A, and denote it as $\mathcal{N}(A)$. It is clear that the set of all solutions of the nonhomogeneous system (5.2) is not a vector space.

Theorem 5.1. The system (5.2) has either a unique solution, no solution, or an infinite number of solutions.

Proof. It suffices to show that if u, v are two different solutions of (5.2), then for any $c \in F$, $u + c(u - v)$ is also a solution of (5.2). But this immediately follows from

$$A[u + c(u - v)] \;=\; Au + c(Au - Av) \;=\; b + c(b - b) \;=\; b. \qquad \blacksquare$$

Related to the system (5.2), the matrix

$$(A|b) \;=\; \left(\begin{array}{cccc|c}
a_{11} & a_{12} & \cdots & a_{1n} & b_1 \\
a_{21} & a_{22} & \cdots & a_{2n} & b_2 \\
\vdots & & & & \\
a_{m1} & a_{m2} & \cdots & a_{mn} & b_m
\end{array} \right) \tag{5.4}$$

is called the *augmented matrix*, which is a partitioned matrix. Clearly, the system (5.2) is completely recognized by its augmented matrix $(A|b)$. In fact, from every augmented matrix of the form (5.4) a corresponding system of the form (5.2) can be written. Therefore, the study of the system (5.2) is equivalent to the study of (5.4). We shall use some elementary operations on $(A|b)$ so that from the reduced matrix the solutions (if any) of the system (5.2) can be obtained rather easily. For this, we begin with the following definition:

An $m \times n$ matrix A is called an *echelon matrix* if all zero rows, if any, appear at the bottom of the matrix, and each leading (first) nonzero element in a row is to the right of the leading nonzero element in the preceding row. Thus, A is an echelon matrix if there exist nonzero elements $a_{1j_1}, a_{2j_2}, \cdots, a_{rj_r}$ where $j_1 < j_2 < \cdots < j_r$, and

$$a_{ij} \;=\; 0 \quad \text{for} \quad \begin{cases} i \le r, \quad j < j_i \\ i > r. \end{cases}$$

These elements $a_{1j_1}, a_{2j_2}, \cdots, a_{rj_r}$ are called the *pivots* of the echelon matrix.

Example 5.1. The following matrix is an echelon matrix

$$A \;=\; \left(\begin{array}{cccccccc}
0 & 3 & 7 & 2 & 5 & 1 & 0 & 9 \\
0 & 0 & 0 & 5 & 3 & 9 & 7 & 1 \\
0 & 0 & 0 & 0 & 0 & 7 & 3 & 5 \\
0 & 0 & 0 & 0 & 0 & 0 & 4 & 8 \\
0 & 0 & 0 & 0 & 0 & 0 & 0 & 7 \\
0 & 0 & 0 & 0 & 0 & 0 & 0 & 0
\end{array} \right)$$

and for this matrix the elements $a_{1j_1} = a_{12} = 3$, $a_{2j_2} = a_{24} = 5$, $a_{3j_3} = a_{36} = 7$, $a_{4j_4} = a_{47} = 4$, $a_{5j_5} = a_{58} = 7$ are the pivots.

Let A be a given $m \times n$ matrix with $R_1, \cdots, R_m$ as rows. The following operations on A are called *elementary row operations*.

1. Interchanging R_i and R_j.
2. Replacing R_i by cR_i, $c \in F$, $c \neq 0$.
3. Replacing R_i by $R_i + cR_j$.

Elementary column operations are defined analogously.

Let I be the $m \times m$ identity matrix. A new matrix obtained from I by employing an elementary row operation is called an *elementary matrix*, and denoted as E. In what follows we shall denote by $E_{ij}, E_i(c), E_{ij}(c)$ the elementary matrices obtained from I by employing, respectively, the above elementary row operations.

Theorem 5.2. An elementary row operation on A is equivalent to EA, where E is the elementary matrix obtained by employing the same elementary row operation.

Proof. We shall prove only for the elementary row operation 3. For this, we note that $E_{ij}(c) = I + c(e^i)^t e^j$, where e^i, $i = 1, \cdots, m$ are the m-tuple unit vectors. Thus, we have

$$E_{ij}(c)A = (I + c(e^i)^t e^j)A = A + c(e^i)^t e^j A.$$

Now it suffices to note that $e^j A = R_j$, and $(e^i)^t R_j$ is the $m \times n$ matrix whose i-th row is R_j, and all other elements are zero. ∎

Theorem 5.3. Each elementary matrix E is nonsingular, and

(i) $E_{ij}^{-1} = E_{ji} = E_{ij}$
(ii) $E_i^{-1}(c) = E_i(1/c)$
(iii) $E_{ij}^{-1}(c) = E_{ij}(-c)$.

Proof. (i) It suffices to notice that $E_{ji}E_{ij} = I$. (ii) Clearly, we have $E_i(c)E_i(1/c) = I$. (iii) As in Theorem 5.2, we find

$$\begin{aligned} E_{ij}(c)E_{ij}(-c) &= (I + c(e^i)^t e^j)(I - c(e^i)^t e^j) \\ &= I + c(e^i)^t e^j - c(e^i)^t e^j - c^2(e^i)^t[e^j(e^i)^t]e^j \\ &= I - c^2(e_i)^t[0]e^j = I. \quad \blacksquare \end{aligned}$$

A matrix B is said to be *row equivalent* to a matrix A, and written as $B \sim A$, if B can be obtained from A by employing a sequence of elementary row operations. It is clear that $\sim$ is an equivalence relation, i.e., 1. $A \sim A$. 2. If $B \sim A$, then $A \sim B$. 3. If $A \sim B$ and $B \sim C$, then $A \sim C$. If B is an echelon matrix, then B is called an *echelon form* of A. *Column equivalent* matrices are defined analogously.

A combination of Theorems 5.2 and 5.3 gives the following result.

Theorem 5.4. If B is row equivalent to a matrix A, then there exists a nonsingular matrix C such that $B = CA$ and $C^{-1}B = A$.

From the above considerations the following results are immediate.

Theorem 5.5. Suppose $A = (a_{ij})$ and $B = (b_{ij})$ are row equivalent echelon matrices with pivots $a_{1j_1}, \cdots, a_{rj_r}$ and $b_{1k_1}, \cdots, b_{sk_s}$, respectively. Then, $r = s$ and $j_i = k_i$, $i = 1, \cdots, r$, i.e., A and B have the same number of nonzero rows and the pivot elements are in the same position.

Theorem 5.6. Every matrix A is row equivalent to an echelon matrix. This matrix is called the *echelon form* of A. However, the echelon form of a matrix A is not unique.

An $m \times n$ matrix A is said to be in *row canonical form* if it is an echelon matrix, each pivot element is equal to 1, and each pivot is the only nonzero element in its column.

Example 5.2. Matrices 0 and I are in row canonical form. The matrix A in Example 5.1 is not in row canonical form, whereas the following matrix is in row canonical form

$$A = \begin{pmatrix} 0 & 1 & 7 & 0 & 5 & 0 & 0 & 9 \\ 0 & 0 & 0 & 1 & 3 & 0 & 0 & 1 \\ 0 & 0 & 0 & 0 & 0 & 1 & 0 & 5 \\ 0 & 0 & 0 & 0 & 0 & 0 & 1 & 8 \\ 0 & 0 & 0 & 0 & 0 & 0 & 0 & 0 \end{pmatrix}.$$

Theorem 5.7. Every matrix A is row equivalent to a unique matrix in row canonical form. This unique matrix is called *row canonical form* of A.

Two linear systems of m equations in n unknowns are said to be *equivalent* if they have the same solution(s). It is clear that an elementary row operation on the augmented matrix $(A|b)$ is equivalent to applying the corresponding operation on the linear system (5.2) itself. Further, the new linear system obtained after applying an elementary row operation on (5.2) is equivalent to the linear system (5.2). Thus, the linear system corresponding to the matrix obtained after applying a finite sequence of row operations on the augmented matrix $(A|b)$ is equivalent to the linear system (5.2). In fact, we have the following result.

Theorem 5.8. The linear system corresponding to the echelon form (row canonical form) known as *echelon linear system* (*row canonical linear system*) of the augmented matrix $(A|b)$ is equivalent to the linear system (5.2).

Theorem 5.9. The linear system (5.2) has a solution if and only if the echelon form (row canonical form) of the augmented matrix $(A|b)$ does not have a row of the form $(0, \cdots, 0, d)$, $d \neq 0$.

Theorem 5.10. For the linear system corresponding to the echelon form (row canonical form) of the augmented matrix $(A|b)$, the following hold:

1. If $r = n$, then the linear system has a unique solution.

2. If $r < n$, then the linear system has an infinite number of solutions. Further, these solutions can be obtained in terms of $(n - r)$ unknowns $\{x_1, \cdots, x_n\} \backslash \{x_{j_1}, \cdots, x_{j_r}\}$.

Theorem 5.11. If $m = n$, the system (5.2) has a unique solution if and only if A^{-1} exists. Further, this solution can be written as $x = A^{-1}b$.

Proof. If (5.2) with $m = n$ has a unique solution, then in view of Theorem 5.10(1) the row canonical form of A is I, and hence from Theorem 5.4 there exists a nonsingular matrix C such that $I = CA$. This implies that A^{-1} exists. Conversely, if A^{-1} exists, then $A(A^{-1}b) = (AA^{-1})b = b$, i.e., $A^{-1}b$ is a solution. To show its uniqueness, let $Au = b$, then $u = (A^{-1}A)u = A^{-1}(Au) = A^{-1}b$. ∎

Corollary 5.1. If $m = n$, the homogeneous system (5.3) has only the trivial solution if and only if $\det(A) \neq 0$. Further, if (5.3) has only the trivial solution, the nonhomogeneous system (5.2) has a unique solution.

Corollary 5.2. The homogeneous system (5.3) with more unknowns than equations has a nonzero solution.

The *rank* of a matrix A, written as rank(A), is equal to the number of pivots r in an echelon form of A; equivalently, A has at least one non-zero minor of order r, and every minor of order larger than r vanishes. From Theorems 5.10 and 5.11, and the fact that an echelon form of the augmented matrix $(A|b)$ automatically yields an echelon form of A, the following result follows.

Theorem 5.12. 1. The system (5.2) has a solution if and only if rank(A) = rank$(A|b)$.

2. The system (5.2) with $m = n$ has a unique solution if and only if rank(A) = rank$(A|b) = n$.

The following example shows how row operations can be used to reduce a matrix to an echelon (row canonical) form.

Example 5.3. Successively, we have

$$A = \begin{pmatrix} 1 & -2 & 3 & 1 \\ 1 & -1 & 4 & 3 \\ 2 & 5 & 7 & 4 \end{pmatrix}_{\substack{R_2 - R_1 \\ R_3 - 2R_1}} \sim \begin{pmatrix} 1 & -2 & 3 & 1 \\ 0 & 1 & 1 & 2 \\ 0 & 9 & 1 & 2 \end{pmatrix}_{R_3 - 9R_2}$$

$$\sim \begin{pmatrix} 1 & -2 & 3 & 1 \\ 0 & 1 & 1 & 2 \\ 0 & 0 & -8 & -16 \end{pmatrix}_{R_3/(-8)} \qquad \text{(echelon form)}$$

$$\sim \begin{pmatrix} 1 & -2 & 3 & 1 \\ 0 & 1 & 1 & 2 \\ 0 & 0 & 1 & 2 \end{pmatrix}_{\substack{R_2 - R_3 \\ R_1 - 3R_3}} \quad \sim \begin{pmatrix} 1 & -2 & 0 & -5 \\ 0 & 1 & 0 & 0 \\ 0 & 0 & 1 & 2 \end{pmatrix} \text{(row canonical form)}.$$

Clearly, for this matrix rank is 3.

Problems

5.1. Prove Theorem 4.1.

5.2. Show that the system (5.2) has a solution if and only if b is a linear combination of the columns of the coefficient matrix A, i.e., b is in the span of the column vectors of A.

5.3. Let $u^1, \cdots, u^m$ and v be vectors in R^n. Show that $v \in \text{Span}\{u^1, \cdots, u^m\}$ if and only if the linear system represented by the augmented matrix $(u^1, \cdots, u^m | v)$ has a solution.

5.4. Show that the system (5.2) is always consistent for at least one vector b.

5.5. Let the system (5.2) with $m = n + 1$ be consistent. Show that $\det(A|b) = 0$; however, the converse is not true. In particular, if the system (5.3) has a nontrivial solution, then $\det(A) = 0$.

5.6. Let the system (5.2) be consistent, and S denote the set of all solutions of this system. Then, $S = u^0 + \mathcal{N}(A)$, where u^0 is a fixed element of S.

5.7. Show that every $m \times n$ matrix A is equivalent to a unique block matrix of the form

$$\begin{pmatrix} I & 0 \\ 0 & 0 \end{pmatrix},$$

where I is the identity matrix of order r.

5.8. Let A and B be $m \times n$ matrices. Show that A is equivalent to B if and only if A^t is equivalent to B^t.

5.9. Find the values of λ for which the system

$$\begin{aligned} (\lambda - 1)x_1 + (3\lambda + 1)x_2 + 2\lambda x_3 &= 0 \\ (\lambda - 1)x_1 + (4\lambda - 2)x_2 + (\lambda + 3)x_3 &= 0 \\ 2x_1 + (3\lambda + 1)x_2 + 3(\lambda - 1)x_3 &= 0 \end{aligned}$$

has a nontrivial solution, and find the ratios of $x : y : z$ when λ has the smallest of these values. What happens when λ has the greatest of these values?

5.10. Find the values of λ and μ so that the system

$$\begin{array}{rcl} 2x_1 + 3x_2 + 5x_3 & = & 9 \\ 7x_1 + 3x_2 - 2x_3 & = & 8 \\ 2x_1 + 3x_2 + \lambda x_3 & = & \mu \end{array}$$

has (i) no solution, (ii) a unique solution, and (iii) an infinite number of solutions.

5.11. Use row operations to reduce the following matrices to echelon (row canonical) form and find their ranks

(i) $\begin{pmatrix} 1 & 3 & 2 \\ 1 & 4 & 2 \\ 2 & 6 & 5 \end{pmatrix}$, (ii) $\begin{pmatrix} 0 & 1 & -3 & -1 \\ 1 & 0 & 1 & 1 \\ 3 & 1 & 0 & 2 \\ 1 & 1 & -2 & 0 \end{pmatrix}$, (iii) $\begin{pmatrix} 1 & -2 & 1 \\ -2 & 5 & -3 \\ 3 & -6 & 3 \\ 1 & -2 & 2 \end{pmatrix}$.

Answers or Hints

5.1. Assume that $BA = I$. If A is invertible, then we can multiply this equation on the right by A^{-1} to get $B = A^{-1}$, from which it follows that $B^{-1} = (A^{-1})^{-1} = A$. This shows that B is invertible, and that A and B are inverses of one another. To show that A is invertible, in view of Corollary 5.1 it suffices to show that the homogeneous system (5.3) with $m = n$ has only the trivial solution. For this, if u is any solution of (5.3), then $BA = I$ implies that $u = Iu = (BA)u = B(Au) = B0 = 0$. The same argument holds when $AB = I$.

5.2. Let $C_1, \cdots, C_n \in R^m$ be the columns of A. Then (5.2) is equivalent to $C_1 x_1 + \cdots + C_n x_n = b$.

5.3. The vector v is in $\mathrm{Span}\{u^1, \cdots, u^m\}$ if and only if there exist scalars $x_1, \cdots, x_m$ such that $x_1 u^1 + \cdots + x_m u^m = v$.

5.4. For $b = 0$ the system (5.2) is always consistent.

5.5. Let $C_1, \cdots, C_n \in R^{n+1}$ be the columns of A. Then $b - C_1 x_1 - \cdots - C_n x_n = 0$. Hence, $\det(A|b) = \det(A|0) = 0$. Consider the system $x + y = 1, 2x + 2y = 3, 3x + 3y = 7$. Let $x_1, \cdots, x_n$ be a nontrivial solution of (5.3). We can assume $x_n \neq 0$. Then the system $a_{i1}(x_1/x_n) + \cdots + a_{i,n-1}(x_{n-1}/x_n) + a_{i,n} = 0$, $i = 1, \cdots, n$ is consistent.

5.6. Let $v \in N(A)$. Then $A(u^0 + v) = Au^0 + Av = b + 0 = b$. Thus $u^0 + v$ is a solution of (5.2). Now suppose that u is a solution of (5.2), then $A(u - u^0) = Au - Au^0 = b - b = 0$. Thus $u - u^0 \in N(A)$. Since $u = u^0 + (u - u^0)$, we find that any solution of (5.2) can be obtained by adding a solution of (5.3) to a fixed solution of (5.2).

5.7. Reduce A to its row canonical form, and then if necessary use elementary column operations.

5.8. Use the definitions of equivalence and transpose of matrices.

5.9. The determinant of the system is $6\lambda(\lambda - 3)^2$. For $\lambda = 0$, $x = y = z$. For $\lambda = 3$ the equations become identical.

5.10. The echelon form of the augmented matrix of the system is

$$\left(\begin{array}{ccc|c} 2 & 3 & 5 & 9 \\ 0 & -15 & -39 & -47 \\ 0 & 0 & \lambda - 5 & \mu - 9 \end{array} \right).$$

(i) $\lambda = 5, \mu \neq 9.$

(ii) $\lambda \neq 5, \mu$ arbitrary.

(iii) $\lambda = 5, \mu = 9.$

5.11. (i) $\begin{pmatrix} 1 & 3 & 2 \\ 0 & 1 & 0 \\ 0 & 0 & 1 \end{pmatrix}$, $\begin{pmatrix} 1 & 0 & 0 \\ 0 & 1 & 0 \\ 0 & 0 & 1 \end{pmatrix}$, 3.

(ii) $\begin{pmatrix} 1 & 0 & 1 & 1 \\ 0 & 1 & -3 & -1 \\ 0 & 0 & 0 & 0 \\ 0 & 0 & 0 & 0 \end{pmatrix}$, 2.

(iii) $\begin{pmatrix} 1 & -2 & 1 \\ 0 & 1 & -1 \\ 0 & 0 & 1 \\ 0 & 0 & 0 \end{pmatrix}$, $\begin{pmatrix} 1 & 0 & 0 \\ 0 & 1 & 0 \\ 0 & 0 & 1 \\ 0 & 0 & 0 \end{pmatrix}$, 3.

Chapter 6

Linear Systems (Cont'd)

We recall that an exact method is an algorithm with a finite and predetermined number of steps, at the end of which it provides a solution of the problem. In this chapter we shall study some exact methods to find the solution(s) of the linear system (5.2). Some of these methods also provide a systematic procedure to compute the value of a determinant, and the inverse of a given matrix.

Cramer's rule. Suppose $m = n$ and the matrix A is nonsingular so that A^{-1} exists. Then, from Theorems 5.11 and 4.5, and the relations (3.1)–(3.3), it follows that

$$x = A^{-1}b = \frac{\text{Adj}\, A}{|A|}b = \frac{(\alpha_{ji})b}{|A|} = \frac{1}{|A|}(|A_1|, |A_2|, \cdots, |A_n|)^t, \quad (6.1)$$

where

$$|A_i| = \sum_{j=1}^{n} \alpha_{ji}b_j = \begin{vmatrix} a_{11} & \cdots & a_{1,i-1} & b_1 & a_{1,i+1} & \cdots & a_{1n} \\ \cdots & & & & & & \\ a_{n1} & \cdots & a_{n,i-1} & b_n & a_{n,i+1} & \cdots & a_{nn} \end{vmatrix}$$

are the determinants obtained from the determinant $|A|$ by replacing its i-th column by the column of constant terms. From (6.1) it follows that

$$x_1 = \frac{|A_1|}{|A|}, \quad x_2 = \frac{|A_2|}{|A|}, \cdots, x_n = \frac{|A_n|}{|A|}. \quad (6.2)$$

Thus to solve the system (5.2) with $m = n$ by Cramer's rule we need to evaluate $(n+1)$ determinants of order n, which is quite a laborious operation, especially when the number n is large. For example, for $n = 10$ we require $359,251,210$ multiplications and divisions if the usual method of expansion of determinants in terms of minors is used. However, Cramer's rule is of theoretical interest.

Example 6.1. To apply Cramer's rule for the system

$$Ax = \begin{pmatrix} 2 & -1 & 1 \\ 3 & 2 & -5 \\ 1 & 3 & -2 \end{pmatrix} \begin{pmatrix} x_1 \\ x_2 \\ x_3 \end{pmatrix} = \begin{matrix} 2x_1 - x_2 + x_3 & = & 0 \\ 3x_1 + 2x_2 - 5x_3 & = & 1 \\ x_1 + 3x_2 - 2x_3 & = & 4 \end{matrix} \quad (6.3)$$

we need to compute

$$|A| = \begin{vmatrix} 2 & -1 & 1 \\ 3 & 2 & -5 \\ 1 & 3 & -2 \end{vmatrix} = 28, \quad |A_1| = \begin{vmatrix} 0 & -1 & 1 \\ 1 & 2 & -5 \\ 4 & 3 & -2 \end{vmatrix} = 13$$

$$|A_2| = \begin{vmatrix} 2 & 0 & 1 \\ 3 & 1 & -5 \\ 1 & 4 & -2 \end{vmatrix} = 47, \quad |A_3| = \begin{vmatrix} 2 & -1 & 0 \\ 3 & 2 & 1 \\ 1 & 3 & 4 \end{vmatrix} = 21.$$

Now (6.2) gives the solution

$$x_1 = \frac{13}{28}, \quad x_2 = \frac{47}{28}, \quad x_3 = \frac{21}{28} = \frac{3}{4}.$$

Gauss elimination method. This is a systematic procedure (*algorithm*) that is often implemented over a machine provided the size of the system (5.1) is not too large. It first reduces the augmented matrix $(A|b)$ with rows $R_1, \cdots, R_m$ to an echelon form, and then suggests how to solve the linear system corresponding to the obtained echelon matrix. We describe the method in the following steps:

1. Find the first column of $(A|b)$ with a nonzero element. Let j_1 denote this column.

2. If necessary, interchange rows of $(A|b)$ so that $|a_{1j_1}| = \max\{|a_{1j_1}|, |a_{2j_1}|, \cdots, |a_{nj_1}|\}$. We once again denote the rows of the rearranged $(A|b)$ by $R_1, \cdots, R_m$.

3. Divide the first row R_1 by the pivot a_{1j_1} and use the resulting row to obtain 0's below a_{1j_1} by replacing R_i, $i = 2, \cdots, m$ by $R_i - (a_{ij_1}/a_{1j_1})R_1$.

4. Consider the matrix obtained in Step 3 excluding the first row. For this submatrix we repeat Steps 1–3. This will give the second pivot a_{2j_2}.

5. Continue Steps 1–4 until all rows (if any) of the submatrix are of the form $(0, \cdots, 0, d)$.

It is clear that Steps 1–5 will reduce $(A|b)$ to an echelon form with the pivots $a_{ij_i} = 1$, $i = 1, \cdots, r \le m$.

6. If the echelon form of $(A|b)$ has a row of the form $(0, \cdots, 0, d)$, $d \ne 0$ the system (5.2) has no solution. Otherwise, the linear system corresponding to the echelon form will be of the form

$$
\begin{aligned}
x_{j_1} + \tilde{a}_{1,j_1+1}x_{j_1+1} + \cdots + \tilde{a}_{1n}x_n &= d_{j_1} \\
&\cdots \\
x_{j_{r-1}} + \tilde{a}_{r-1,j_{r-1}+1}x_{j_{r-1}+1} + \cdots + \tilde{a}_{r-1,n}x_n &= d_{j_{r-1}} \\
x_{j_r} + \tilde{a}_{r,j_r+1}x_{j_r+1} + \cdots + \tilde{a}_{r,n}x_n &= d_{j_r}.
\end{aligned}
\tag{6.4}
$$

If $r < n$, we compute x_{j_i}, $i = 1, \cdots, r$ (known as *basic variables*) in terms of $(n-r)$ unknowns $\{x_1, \cdots, x_n\} \setminus \{x_{j_1}, \cdots, x_{j_r}\}$ (called *free variables*) in the reverse order (*back substitution*) by the formulae

$$x_{j_i} = d_{j_i} - \sum_{j=j_i+1}^{n} \tilde{a}_{ij} x_j, \quad i = r, r-1, \cdots, 1. \tag{6.5}$$

In particular, if $r = n$, then $j_i = i$, $i = 1, \cdots, n$ and the formulae reduce to

$$x_n = d_n, \quad x_i = d_i - \sum_{j=i+1}^{n} \tilde{a}_{ij} x_j, \quad i = n-1, n-2, \cdots, 1. \tag{6.6}$$

Remark 6.1. To solve the system (5.2) with $m = n$ by the Gauss elimination method, we need a total number of

$$\frac{n}{3}(n^2 + 3n - 1) = \mathcal{O}\left(\frac{n^3}{3}\right) \tag{6.7}$$

multiplications and divisions. In particular, for $n = 10$ this number is 430, which is very tiny compared to what we need by Cramer's rule.

Example 6.2. For the system (6.3), we have

$$(A|b) = \begin{pmatrix} 2 & -1 & 1 & | & 0 \\ 3 & 2 & -5 & | & 1 \\ 1 & 3 & -2 & | & 4 \end{pmatrix} \sim \begin{pmatrix} 3 & 2 & -5 & | & 1 \\ 2 & -1 & 1 & | & 0 \\ 1 & 3 & -2 & | & 4 \end{pmatrix}$$

$$\sim \begin{pmatrix} 1 & \frac{2}{3} & -\frac{5}{3} & | & \frac{1}{3} \\ 0 & -\frac{7}{3} & \frac{13}{3} & | & -\frac{2}{3} \\ 0 & \frac{7}{3} & -\frac{1}{3} & | & \frac{11}{3} \end{pmatrix} \sim \begin{pmatrix} 1 & \frac{2}{3} & -\frac{5}{3} & | & \frac{1}{3} \\ 0 & 1 & -\frac{13}{7} & | & \frac{2}{7} \\ 0 & 0 & 4 & | & 3 \end{pmatrix}$$

$$\sim \begin{pmatrix} 1 & \frac{2}{3} & -\frac{5}{3} & | & \frac{1}{3} \\ 0 & 1 & -\frac{13}{7} & | & \frac{2}{7} \\ 0 & 0 & 1 & | & \frac{3}{4} \end{pmatrix}$$

and hence, from (6.6), we find

$$x_3 = \frac{3}{4}, \quad x_2 = \frac{2}{7} + \frac{13}{7} \times \frac{3}{4} = \frac{47}{28}, \quad x_1 = \frac{1}{3} - \frac{2}{3} \times \frac{47}{28} + \frac{5}{3} \times \frac{3}{4} = \frac{13}{28}.$$

Example 6.3. For the system

$$\begin{aligned} 2x_1 + x_2 + x_3 &= 1 \\ x_1 + 2x_2 + x_3 &= 2 \\ x_1 + x_2 + 2x_3 &= 3 \\ 5x_1 + 4x_2 + 3x_3 &= 8 \end{aligned}$$

we have

$$(A|b) = \begin{pmatrix} 2 & 1 & 1 & 1 \\ 1 & 2 & 1 & 2 \\ 1 & 1 & 2 & 3 \\ 5 & 4 & 3 & 8 \end{pmatrix} \sim \begin{pmatrix} 1 & \frac{1}{2} & \frac{1}{2} & \frac{1}{2} \\ 0 & 1 & \frac{1}{3} & 1 \\ 0 & 0 & 1 & \frac{3}{2} \\ 0 & 0 & 0 & 4 \end{pmatrix}$$

and hence it has no solution.

Example 6.4. For the system

$$2x_1 + x_2 + x_3 = 1$$
$$x_1 + 2x_2 + x_3 = 2$$

we have

$$(A|b) = \begin{pmatrix} 2 & 1 & 1 & 1 \\ 1 & 2 & 1 & 2 \end{pmatrix} \sim \begin{pmatrix} 1 & \frac{1}{2} & \frac{1}{2} & \frac{1}{2} \\ 0 & 1 & \frac{1}{3} & 1 \end{pmatrix}$$

and hence it has an infinite number of solutions $x_1 = -(1/3)x_3$, $x_2 = 1 - (1/3)x_3$, where x_3 is arbitrary.

The Gauss elimination method can be used to compute the determinant of a given $n \times n$ matrix A. If all $a_{i1} = 0$, $i = 1, \cdots, n$ then $\det(A) = 0$; otherwise, if necessary we interchange rows of A so that $|a_{11}| = \max\{|a_{11}|, |a_{21}|, \cdots, |a_{n1}|\}$. Now following Step 3 with $j_1 = 1$, we get

$$|A| = \theta_1 a_{11} \begin{vmatrix} 1 & \alpha_{12} & \cdots & \alpha_{1n} \\ 0 & \alpha_{22} & \cdots & \alpha_{2n} \\ & \cdots & & \\ 0 & \alpha_{n2} & \cdots & \alpha_{nn} \end{vmatrix} = \theta_1 a_{11} \begin{vmatrix} \alpha_{22} & \cdots & \alpha_{2n} \\ \cdots & & \\ \alpha_{n2} & \cdots & \alpha_{nn} \end{vmatrix},$$

where θ_1 is -1 or 1 accordingly as the number of rows interchanged is odd or even, $\alpha_{1j} = a_{1j}/a_{11}$, $2 \le j \le n$, $\alpha_{ij} = a_{ij} - \alpha_{1j}a_{i1}$, $2 \le i, j \le n$. Thus, we could reduce the order of the determinant from n to $n-1$. We continue this process $n-1$ times.

Remark 6.2. The total number of multiplications and divisions necessary for evaluating the determinant $|A|$ by Gauss elimination technique is

$$\frac{(n-1)}{3}(n^2 + n + 3) = \mathcal{O}\left(\frac{n^3}{3}\right). \tag{6.8}$$

Hence, if we compute all $(n+1)$ required determinants in Cramer's rule by the Gauss elimination method, we will need

$$\frac{(n-1)}{3}(n^2 + n + 3)(n+1) = \mathcal{O}\left(\frac{n^4}{3}\right) \tag{6.9}$$

as the number of multiplications and divisions.

Example 6.5. For computing the determinant of the matrix A in (6.3), successively, we have

$$
|A| = (-1) \begin{vmatrix} 3 & 2 & -5 \\ 2 & -1 & 1 \\ 1 & 3 & -2 \end{vmatrix} = (-1)(3) \begin{vmatrix} 1 & \frac{2}{3} & -\frac{5}{3} \\ 2 & -1 & 1 \\ 1 & 3 & -2 \end{vmatrix}
$$

$$
= (-1)(3) \begin{vmatrix} 1 & \frac{2}{3} & -\frac{5}{3} \\ 0 & -\frac{7}{3} & \frac{13}{3} \\ 0 & \frac{7}{3} & -\frac{1}{3} \end{vmatrix} = (-1)(3) \begin{vmatrix} -\frac{7}{3} & \frac{13}{3} \\ \frac{7}{3} & -\frac{1}{3} \end{vmatrix}
$$

$$
= (-1)(3)\left(-\frac{7}{3}\right) \begin{vmatrix} 1 & -\frac{13}{7} \\ \frac{7}{3} & -\frac{1}{3} \end{vmatrix} = (-1)(3)\left(-\frac{7}{3}\right) \begin{vmatrix} 1 & -\frac{13}{7} \\ 0 & 4 \end{vmatrix}
$$

$$
= (-1)(3)\left(-\frac{7}{3}\right)(4) = 28.
$$

Systems of the form $AX = B$, where $A = (a_{ij})$ is an $n \times n$ matrix, and $B = (b_{ij})$, $X = (x_{ij})$ are $n \times p$ matrices, can also be solved simultaneously by the Gauss elimination technique. In fact, if we write the augmented matrix of the elements of A and B, i.e.,

$$
(A|B) = \begin{pmatrix} a_{11} & \cdots & a_{1n} & b_{11} & \cdots & b_{1p} \\ a_{21} & \cdots & a_{2n} & b_{21} & \cdots & b_{2p} \\ \cdots & & & & & \\ a_{n1} & \cdots & a_{nn} & b_{n1} & \cdots & b_{np} \end{pmatrix}
$$

and assume that $\det(A) \neq 0$, then the Gauss elimination process gives

$$
\begin{pmatrix} 1 & c_{12} & \cdots & c_{1,n-1} & c_{1n} & d_{11} & \cdots & d_{1p} \\ & 1 & \cdots & c_{2,n-1} & c_{2n} & d_{21} & \cdots & d_{2p} \\ & & \cdots & & & & & \\ & & & 1 & c_{n-1,n} & d_{n-1,1} & \cdots & d_{n-1,p} \\ & & & & 1 & d_{n1} & \cdots & d_{np} \end{pmatrix}.
$$

The unknowns x_{ik}, $k = 1, 2, \cdots, p$ are now obtained from

$$
x_{nk} = d_{nk}, \quad x_{ik} = d_{ik} - \sum_{j=i+1}^{n} c_{ij} x_{jk}, \quad i = n-1, n-2, \cdots, 1. \quad (6.10)
$$

In particular, if we consider the system $AX = I$, where $X = (x_{ij})$ is an $n \times n$ matrix and I is the $n \times n$ identity matrix, then since $X = A^{-1}$, the matrix (x_{ik}) formed from (6.10) gives the inverse of the matrix A.

Example 6.6. For computing the inverse of the matrix A in (6.3), successively, we have

$$\begin{pmatrix} 2 & -1 & 1 & | & 1 & 0 & 0 \\ 3 & 2 & -5 & | & 0 & 1 & 0 \\ 1 & 3 & -2 & | & 0 & 0 & 1 \end{pmatrix} \sim \begin{pmatrix} 3 & 2 & -5 & | & 0 & 1 & 0 \\ 2 & -1 & 1 & | & 1 & 0 & 0 \\ 1 & 3 & -2 & | & 0 & 0 & 1 \end{pmatrix}$$

$$\sim \begin{pmatrix} 1 & \frac{2}{3} & -\frac{5}{3} & | & 0 & \frac{1}{3} & 0 \\ 2 & -1 & 1 & | & 1 & 0 & 0 \\ 1 & 3 & -2 & | & 0 & 0 & 1 \end{pmatrix} \sim \begin{pmatrix} 1 & \frac{2}{3} & -\frac{5}{3} & | & 0 & \frac{1}{3} & 0 \\ 0 & -\frac{7}{3} & \frac{13}{3} & | & 1 & -\frac{2}{3} & 0 \\ 0 & \frac{7}{3} & -\frac{1}{3} & | & 0 & -\frac{1}{3} & 1 \end{pmatrix}$$

$$(6.11)$$

$$\sim \begin{pmatrix} 1 & \frac{2}{3} & -\frac{5}{3} & | & 0 & \frac{1}{3} & 0 \\ 0 & 1 & -\frac{13}{7} & | & -\frac{3}{7} & \frac{2}{7} & 0 \\ 0 & 0 & 4 & | & 1 & -1 & 1 \end{pmatrix} \sim \begin{pmatrix} 1 & \frac{2}{3} & -\frac{5}{3} & | & 0 & \frac{1}{3} & 0 \\ 0 & 1 & -\frac{13}{7} & | & -\frac{3}{7} & \frac{2}{7} & 0 \\ 0 & 0 & 1 & | & \frac{1}{4} & -\frac{1}{4} & \frac{1}{4} \end{pmatrix}.$$

Now (6.10) gives

$$A^{-1} = \begin{pmatrix} \frac{11}{28} & \frac{1}{28} & \frac{3}{28} \\ \frac{1}{28} & -\frac{5}{28} & \frac{13}{28} \\ \frac{1}{4} & -\frac{1}{4} & \frac{1}{4} \end{pmatrix}.$$

Gauss–Jordan elimination method. In the Gauss elimination method, 0's are obtained not only below the pivots a_{ij_i}, $i = 1, 2, \cdots, r$ but above them also, so that in the final reduced matrix each pivot element is equal to 1 and each pivot is the only nonzero element in its column. Thus the matrix $(A|b)$ is reduced to row canonical form. Now the solution can be obtained without using the back substitution. However, then, for the case $m = n$, the total number of multiplications and divisions required is

$$\frac{n^2}{2}(n+1) = \mathcal{O}\left(\frac{n^3}{2}\right). \tag{6.12}$$

Hence, the Gauss–Jordan elimination method is more expansive than the Gauss elimination method. However, it gives a simpler procedure for computing the inverse of a square matrix A. We simply write the augmented matrix of A and I, i.e., $(A|I)$, and when the Gauss–Jordan procedure is complete, we obtain $(I|A^{-1})$.

Example 6.7. In Example 6.6 we follow up to (6.11), and then we have

$$\sim \begin{pmatrix} 1 & 0 & -\frac{3}{7} & | & \frac{2}{7} & \frac{1}{7} & 0 \\ 0 & 1 & -\frac{13}{7} & | & -\frac{3}{7} & \frac{2}{7} & 0 \\ 0 & 0 & 1 & | & \frac{1}{4} & -\frac{1}{4} & \frac{1}{4} \end{pmatrix} \sim \begin{pmatrix} 1 & 0 & 0 & | & \frac{11}{28} & \frac{1}{28} & \frac{3}{28} \\ 0 & 1 & 0 & | & \frac{1}{28} & -\frac{5}{28} & \frac{13}{28} \\ 0 & 0 & 1 & | & \frac{1}{4} & -\frac{1}{4} & \frac{1}{4} \end{pmatrix}.$$

Problems

6.1. Find the echelon form of the given matrix A, and find its null space $\mathcal{N}(A)$

(i) $\quad A = \begin{pmatrix} 1 & 3 & 3 & 4 \\ 2 & 6 & 9 & 9 \\ 3 & 9 & 12 & 16 \end{pmatrix}$, (ii) $A = \begin{pmatrix} 2 & 4 & 0 & 2 & 3 \\ 0 & 3 & 3 & 1 & 5 \\ 2 & 7 & 9 & 7 & 2 \\ 0 & 0 & 6 & 5 & 3 \end{pmatrix}$

(iii) $\quad A = \begin{pmatrix} 0 & 1 & 2 & 0 & 3 & 6 \\ 1 & 2 & 5 & 3 & 4 & 3 \\ 1 & 2 & 5 & 6 & 3 & 4 \end{pmatrix}$.

6.2. Solve the system (5.2) by Cramer's rule, Gauss elimination method, and Gauss–Jordan elimination method, when

(i) $\quad A = \begin{pmatrix} 2 & 1 & -5 & 1 \\ 1 & -3 & 0 & -6 \\ 0 & 2 & -1 & 2 \\ 1 & 4 & -7 & 6 \end{pmatrix}$, $b = \begin{pmatrix} 1 \\ -7 \\ 3 \\ 5 \end{pmatrix}$

(ii) $\quad A = \begin{pmatrix} 2 & 1 & 3 & 1 \\ 4 & 2 & 2 & -2 \\ -2 & -4 & 4 & -3 \\ -4 & 1 & 0 & -1 \end{pmatrix}$, $b = \begin{pmatrix} 17 \\ 6 \\ -10 \\ -6 \end{pmatrix}$

(iii) $\quad A = \begin{pmatrix} 2 & 1 & 0 & 1 \\ 5 & -4 & 1 & 0 \\ 3 & 0 & 2 & 0 \\ 1 & 1 & -1 & 1 \end{pmatrix}$, $b = \begin{pmatrix} 2 \\ 1 \\ -2 \\ 1 \end{pmatrix}$

(iv) $\quad A = \begin{pmatrix} -4 & 10 & -10 & -4 \\ -3 & -5 & -4 & 9 \\ -3 & -9 & -1 & 6 \\ 8 & 1 & 1 & 2 \end{pmatrix}$, $b = \begin{pmatrix} 78 \\ -22 \\ -31 \\ -37 \end{pmatrix}$.

6.3. Use the Gauss elimination method to find the solutions, if any, of the system (5.2) when

(i) $\quad A = \begin{pmatrix} 2 & 7 & 4 & 3 \\ 8 & 5 & 3 & 9 \\ 1 & 3 & 6 & 4 \end{pmatrix}$, $b = \begin{pmatrix} 1 \\ 3 \\ 7 \end{pmatrix}$

(ii) $\quad A = \begin{pmatrix} 5 & 3 & 8 & 1 \\ 6 & 3 & 8 & 6 \\ 4 & 8 & 3 & 5 \end{pmatrix}$, $b = \begin{pmatrix} 3 \\ 5 \\ 7 \end{pmatrix}$

(iii) $A = \begin{pmatrix} 1 & 1 & -3 \\ 2 & 1 & -2 \\ 1 & 1 & 1 \\ 1 & 2 & -3 \end{pmatrix}$, $\quad b = \begin{pmatrix} -1 \\ 1 \\ 3 \\ 1 \end{pmatrix}$

(iv) $A = \begin{pmatrix} 2 & 1 & 1 \\ 1 & 3 & 1 \\ 1 & 1 & 5 \\ 2 & 3 & -3 \end{pmatrix}$, $\quad b = \begin{pmatrix} 2 \\ 5 \\ -7 \\ 14 \end{pmatrix}$

(v) $A = \begin{pmatrix} 4 & 2 & -1 & 1 & -1 \\ 1 & -3 & 2 & 1 & 1 \\ 3 & -1 & 3 & -1 & -1 \end{pmatrix}$, $\quad b = \begin{pmatrix} 1 \\ -1 \\ 2 \end{pmatrix}$.

6.4. Use the Gauss elimination method to find all solutions of the homogeneous system (5.3) when the matrix A is as in Problem 6.3.

6.5. Use the Gauss elimination method to find the value of the following determinants:

(i) $\begin{vmatrix} 1 & -1 & 3 & 5 \\ 3 & 7 & -8 & -4 \\ 3 & 9 & 2 & 0 \\ 3 & 0 & 5 & -3 \end{vmatrix}$, (ii) $\begin{vmatrix} 27 & 13 & 19 & 17 \\ -5 & -19 & -29 & -31 \\ 11 & -13 & 17 & 29 \\ 3 & 0 & 25 & 4 \end{vmatrix}$.

6.6. Compute the inverse of the following matrices by using the Gauss elimination method, and the Gauss–Jordan elimination method:

(i) $\begin{pmatrix} 0 & 1 & 2 \\ 1 & 2 & 3 \\ 3 & 1 & 1 \end{pmatrix}$, (ii) $\begin{pmatrix} 7 & 6 & 2 \\ -1 & 2 & 4 \\ 3 & 6 & 8 \end{pmatrix}$

(iii) $\begin{pmatrix} 8 & 4 & -3 \\ 2 & 1 & 1 \\ 1 & 2 & 1 \end{pmatrix}$, (iv) $\begin{pmatrix} 1 & 1 & 1 & 1 \\ 1 & 1 & 1 & -4 \\ 1 & 1 & -1 & 1 \\ 1 & -1 & 1 & 1 \end{pmatrix}$.

Answers or Hints

6.1. (i) $\begin{pmatrix} 1 & 3 & 3 & 4 \\ 0 & 0 & 3 & 1 \\ 0 & 0 & 1 & 3 \end{pmatrix}$, $\mathcal{N}(A) = \{(-3,1,0,0)^t\}$.

(ii) $\begin{pmatrix} 2 & 4 & 0 & 2 & 3 \\ 0 & 3 & 3 & 1 & 5 \\ 0 & 0 & 6 & 4 & -6 \\ 0 & 0 & 0 & 1 & 9 \end{pmatrix}$, $\mathcal{N}(A) = \{(\frac{113}{6}, -\frac{17}{3}, 7, -9, 1)^t\}$.

(iii) $\begin{pmatrix} 1 & 2 & 5 & 3 & 4 & 3 \\ 0 & 1 & 2 & 0 & 3 & 6 \\ 0 & 0 & 0 & 3 & -1 & 1 \end{pmatrix}$,

$\mathcal{N}(A) = \{(-1, -2, 1, 0, 0, 0)^t, (1, -3, 0, 1/3, 1, 0)^t, (10, -6, 0, -1/3, 0, 1)^t\}$.

6.2. (i) $x_1 = 2$, $x_2 = 1$, $x_3 = 1$, $x_4 = 1$.

(ii) $x_1 = 1$, $x_2 = 2$, $x_3 = 3$, $x_4 = 4$.

(iii) $x_1 = -4$, $x_2 = -4$, $x_3 = 5$, $x_4 = 14$.

(iv) $x_1 = -4$, $x_2 = 3$, $x_3 = -2$, $x_4 = -3$.

6.3. (i) $\left(\frac{62}{197}, -\frac{161}{197}, \frac{300}{197}, 0\right)^t + \left(-\frac{203}{197}, \frac{41}{197}, -\frac{118}{197}, 1\right)^t c.$

(ii) $\left(2, \frac{13}{55}, -\frac{53}{55}, 0\right)^t + \left(-5, \frac{48}{55}, \frac{147}{55}, 1\right)^t c.$

(iii) No solution.

(iv) $(1, 2, -2)^t.$

(v) $(0, 1, 1, 0, 0)^t + \left(-\frac{13}{30}, \frac{9}{10}, \frac{16}{15}, 1, 0\right)^t c + \left(\frac{1}{30}, \frac{7}{10}, \frac{8}{15}, 0, 1\right)^t d.$

6.4. (i) $\left(-\frac{203}{197}, \frac{41}{197}, -\frac{118}{197}, 1\right)^t c.$

(ii) $\left(-5, \frac{48}{55}, \frac{147}{55}, 1\right)^t c.$

(iii) $(0, 0, 0)^t.$

(iv) $(0, 0, 0)^t.$

(v) $\left(-\frac{13}{30}, \frac{9}{10}, \frac{16}{15}, 1, 0\right)^t c + \left(\frac{1}{30}, \frac{7}{10}, \frac{8}{15}, 0, 1\right)^t d.$

6.5. (i) -1734

(ii) 517824.

6.6. (i) $\begin{pmatrix} \frac{1}{2} & -\frac{1}{2} & \frac{1}{2} \\ -4 & 3 & -1 \\ \frac{5}{2} & -\frac{3}{2} & \frac{1}{2} \end{pmatrix}$

(ii) $\begin{pmatrix} -\frac{1}{5} & -\frac{9}{10} & \frac{1}{2} \\ \frac{1}{2} & \frac{5}{4} & -\frac{3}{4} \\ -\frac{3}{10} & -\frac{3}{5} & \frac{1}{2} \end{pmatrix}$

(iii) $\begin{pmatrix} \frac{1}{21} & \frac{10}{21} & -\frac{1}{3} \\ \frac{1}{21} & -\frac{11}{21} & \frac{2}{3} \\ -\frac{1}{7} & \frac{4}{7} & 0 \end{pmatrix}$

(iv) $\begin{pmatrix} -\frac{1}{5} & \frac{1}{5} & \frac{1}{2} & \frac{1}{2} \\ \frac{1}{2} & 0 & 0 & -\frac{1}{2} \\ \frac{1}{2} & 0 & -\frac{1}{2} & 0 \\ \frac{1}{5} & -\frac{1}{5} & 0 & 0 \end{pmatrix}.$

Chapter 7

Factorization

In this chapter we shall discuss a modified but restricted realization of Gaussian elimination. It factorizes a given $m \times n$ matrix A to a product of two matrices L and U, where L is an $m \times m$ lower triangular matrix, and U is an $m \times n$ upper triangular matrix. We shall also discuss various variants and applications of this factorization.

Theorem 7.1 (LU factorization). Let A be an $m \times n$ matrix. If A can be reduced to echelon form U without interchanging rows, then there exists an $m \times m$ lower triangular matrix L with 1's on the diagonal such that $A = LU$.

Proof. Let $E_1, E_2, \cdots, E_r$ be the $m \times m$ elementary matrices that correspond to the row operations which are needed to reduce A to the echelon form U, i.e.,

$$(E_r \cdots E_2 E_1)A = U. \tag{7.1}$$

Notice that during our reduction of A to U we are allowed only to add a multiple of one row to a row below. This means each row operation E_i corresponds to an $m \times m$ elementary lower triangular matrix with 1's on the diagonal. Now from Problems 2.2 and 4.4 it follows that both $(E_r \cdots E_2 E_1)$ and $(E_r \cdots E_2 E_1)^{-1}$ are $m \times m$ lower triangular matrix with 1's on the diagonal. Thus, from (7.1), we have $A = LU$, where

$$L = (E_r \cdots E_2 E_1)^{-1} = E_1^{-1} E_2^{-1} \cdots E_r^{-1}. \quad \blacksquare \tag{7.2}$$

Example 7.1. Successively, we have

$$A = \begin{pmatrix} 3 & 2 & 3 & 1 \\ 3 & 1 & 4 & 3 \\ -6 & 4 & 2 & 1 \end{pmatrix}, \quad L = \begin{pmatrix} 1 & 0 & 0 \\ \bullet & 1 & 0 \\ \bullet & \bullet & 1 \end{pmatrix}$$

$$R_2 - R_1, R_3 + 2R_1$$

$$A \sim \begin{pmatrix} 3 & 2 & 3 & 1 \\ 0 & -1 & 1 & 2 \\ 0 & 8 & 8 & 3 \end{pmatrix}, \quad L = \begin{pmatrix} 1 & 0 & 0 \\ 1 & 1 & 0 \\ -2 & \bullet & 1 \end{pmatrix}$$

$$R_3 + 8R_2$$

$$U = \begin{pmatrix} 3 & 2 & 3 & 1 \\ 0 & -1 & 1 & 2 \\ 0 & 0 & 16 & 19 \end{pmatrix}, \quad L = \begin{pmatrix} 1 & 0 & 0 \\ 1 & 1 & 0 \\ -2 & -8 & 1 \end{pmatrix}.$$

Example 7.2. Successively, we have

$$A = \begin{pmatrix} 2 & 1 & 3 \\ 4 & 3 & 5 \\ 8 & 2 & 16 \\ -4 & 1 & -11 \\ -2 & 3 & -5 \end{pmatrix}, \quad L = \begin{pmatrix} 1 & 0 & 0 & 0 & 0 \\ \bullet & 1 & 0 & 0 & 0 \\ \bullet & \bullet & 1 & 0 & 0 \\ \bullet & \bullet & \bullet & 1 & 0 \\ \bullet & \bullet & \bullet & \bullet & 1 \end{pmatrix}$$

$$R_2 - 2R_1, R_3 - 4R_1, R_4 + 2R_1, R_5 + R_1$$

$$A \sim \begin{pmatrix} 2 & 1 & 3 \\ 0 & 1 & -1 \\ 0 & -2 & 4 \\ 0 & 3 & -5 \\ 0 & 4 & -2 \end{pmatrix}, \quad L = \begin{pmatrix} 1 & 0 & 0 & 0 & 0 \\ 2 & 1 & 0 & 0 & 0 \\ 4 & \bullet & 1 & 0 & 0 \\ -2 & \bullet & \bullet & 1 & 0 \\ -1 & \bullet & \bullet & \bullet & 1 \end{pmatrix}$$

$$R_3 + 2R_2, R_4 - 3R_2, R_5 - 4R_2$$

$$A \sim \begin{pmatrix} 2 & 1 & 3 \\ 0 & 1 & -1 \\ 0 & 0 & 2 \\ 0 & 0 & -2 \\ 0 & 0 & 2 \end{pmatrix}, \quad L = \begin{pmatrix} 1 & 0 & 0 & 0 & 0 \\ 2 & 1 & 0 & 0 & 0 \\ 4 & -2 & 1 & 0 & 0 \\ -2 & 3 & \bullet & 1 & 0 \\ -1 & 4 & \bullet & \bullet & 1 \end{pmatrix}$$

$$R_4 + R_3, R_5 - R_3$$

$$U = \begin{pmatrix} 2 & 1 & 3 \\ 0 & 1 & -1 \\ 0 & 0 & 2 \\ 0 & 0 & 0 \\ 0 & 0 & 0 \end{pmatrix}, \quad L = \begin{pmatrix} 1 & 0 & 0 & 0 & 0 \\ 2 & 1 & 0 & 0 & 0 \\ 4 & -2 & 1 & 0 & 0 \\ -2 & 3 & -1 & 1 & 0 \\ -1 & 4 & 1 & \bullet & 1 \end{pmatrix}.$$

In L the remaining $\bullet$ is finally replaced by 0.

Remark 7.1 ($LD\hat{U}$ factorization). From the matrix U in Theorem 7.1, we can factor out an $m \times m$ diagonal matrix D so that $U = D\hat{U}$, where the nonzero elements in U and $\hat{U}$ are at the same position; however, in $\hat{U}$ all pivot elements are 1. Clearly, with such an adjustment $A = LD\hat{U}$.

Example 7.3. For the matrix in Example 4.2, we have

$$A = \begin{pmatrix} 2 & -1 & 1 \\ 3 & 2 & -5 \\ 1 & 3 & -2 \end{pmatrix} = \begin{pmatrix} 1 & 0 & 0 \\ \frac{3}{2} & 1 & 0 \\ \frac{1}{2} & 1 & 1 \end{pmatrix} \begin{pmatrix} 2 & -1 & 1 \\ 0 & \frac{7}{2} & -\frac{13}{2} \\ 0 & 0 & 4 \end{pmatrix}$$

$$= \begin{pmatrix} 1 & 0 & 0 \\ \frac{3}{2} & 1 & 0 \\ \frac{1}{2} & 1 & 1 \end{pmatrix} \begin{pmatrix} 2 & 0 & 0 \\ 0 & \frac{7}{2} & 0 \\ 0 & 0 & 4 \end{pmatrix} \begin{pmatrix} 1 & -\frac{1}{2} & \frac{1}{2} \\ 0 & 1 & -\frac{13}{7} \\ 0 & 0 & 1 \end{pmatrix}.$$

Thus, $\det(A) = 1 \times \left(2 \times \frac{7}{2} \times 4\right) \times 1 = 28$.

Example 7.4. The matrix U in Example 7.1 can be written as $U = D\hat{U}$, where

$$D = \begin{pmatrix} 3 & 0 & 0 \\ 0 & -1 & 0 \\ 0 & 0 & 16 \end{pmatrix} \quad \text{and} \quad \hat{U} = \begin{pmatrix} 1 & \frac{2}{3} & 1 & \frac{1}{3} \\ 0 & 1 & -1 & -2 \\ 0 & 0 & 1 & \frac{19}{16} \end{pmatrix}.$$

Remark 7.2 (*PLU* factorization). If $A = PB$, where P is an $m \times m$ permutation matrix, and B can be reduced to echelon form without interchanging rows, then there exists an $m \times m$ lower triangular matrix L with 1's on the diagonal such that $A = PLU$.

Example 7.5. For the matrix in Problem 5.11(ii), we have

$$A = \begin{pmatrix} 0 & 1 & -3 & -1 \\ 1 & 0 & 1 & 1 \\ 3 & 1 & 0 & 2 \\ 1 & 1 & -2 & 0 \end{pmatrix} = \begin{pmatrix} 0 & 1 & 0 & 0 \\ 1 & 0 & 0 & 0 \\ 0 & 0 & 1 & 0 \\ 0 & 0 & 0 & 1 \end{pmatrix} \begin{pmatrix} 1 & 0 & 1 & 1 \\ 0 & 1 & -3 & -1 \\ 3 & 1 & 0 & 2 \\ 1 & 1 & -2 & 0 \end{pmatrix}$$

$$= \begin{pmatrix} 0 & 1 & 0 & 0 \\ 1 & 0 & 0 & 0 \\ 0 & 0 & 1 & 0 \\ 0 & 0 & 0 & 1 \end{pmatrix} \begin{pmatrix} 1 & 0 & 0 & 0 \\ 0 & 1 & 0 & 0 \\ 3 & 1 & 1 & 0 \\ 1 & 1 & 0 & 1 \end{pmatrix} \begin{pmatrix} 1 & 0 & 1 & 1 \\ 0 & 1 & -3 & -1 \\ 0 & 0 & 0 & 0 \\ 0 & 0 & 0 & 0 \end{pmatrix}.$$

From this factorization it immediately follows that $\det(A) = -1 \times 1 \times 0 = 0$.

If conditions of Theorem 7.1 are satisfied, then the system (5.2) can be written as

$$LUx = b. \tag{7.3}$$

In (7.3), we let

$$Ux = y \tag{7.4}$$

so that (7.3) becomes

$$Ly = b. \tag{7.5}$$

Thus, solving (5.2) is equivalent to finding solutions of two simplified systems, namely, first (7.5) to obtain y, and then (7.4) to find x. Since L is a lower triangular matrix with 1's on the diagonal, in expended form the system (7.5) is of the form

$$\ell_{i1}y_1 + \ell_{i2}y_2 + \cdots + \ell_{i,i-1}y_{i-1} + y_i = b_i, \quad i = 1, \cdots, m.$$

Clearly, this system can be solved recursively (forward substitution), to obtain

$$y_i = b_i - \sum_{j=1}^{i-1} \ell_{ij} y_j, \quad i = 1, \cdots, m. \tag{7.6}$$

Example 7.6. Consider the system

$$A(x_1, x_2, x_3, x_4)^t = (1, 2, -9)^t, \tag{7.7}$$

where the matrix A is the same as in Example 7.1. For (7.7), the system (7.6) is

$$
\begin{array}{rclcrcl}
y_1 & = & 1 & & y_1 & = & 1 \\
y_2 & = & 2 - y_1 & \Longrightarrow & y_2 & = & 1 \\
y_3 & = & -9 + 2y_1 + 8y_2 & & y_3 & = & 1.
\end{array}
$$

Now the system (7.4) can be written as

$$
\begin{array}{rcl}
3x_1 + 2x_2 + 3x_3 + x_4 & = & 1 \\
-x_2 + x_3 + 2x_4 & = & 1 \\
16x_3 + 19x_4 & = & 1,
\end{array}
$$

which from the backward substitution gives the solution of (7.7),

$$(x_1, x_2, x_3, x_4)^t = \frac{1}{48}(43, -45, 3, 0)^t + (15, 39, -57, 48)^t c.$$

Example 7.7. Consider the system

$$A(x_1, x_2, x_3)^t = (1, 2, 5, -3, 0)^t, \tag{7.8}$$

where the matrix A is the same as in Example 7.2. For (7.8), the system (7.6) is

$$
\begin{array}{rclcrcl}
y_1 & = & 1 & & y_1 & = & 1 \\
y_2 & = & 2 - 2y_1 & & y_2 & = & 0 \\
y_3 & = & 5 - 4y_1 + 2y_2 & \Longrightarrow & y_3 & = & 1 \\
y_4 & = & -3 + 2y_1 - 3y_2 + y_3 & & y_4 & = & 0 \\
y_5 & = & y_1 - 4y_2 - y_3 & & y_5 & = & 0.
\end{array}
$$

Now the system (7.4) can be written as

$$
\begin{array}{rcl}
2x_1 + x_2 + 3x_3 & = & 1 \\
x_2 - x_3 & = & 0 \\
2x_3 & = & 1,
\end{array}
$$

which from the backward substitution gives the solution of (7.8),

$$(x_1, x_2, x_3)^t = \frac{1}{2}(-1, 1, 1)^t.$$

Remark 7.3. To solve the system (5.2), Gaussian elimination as well as LU factorization require the same amount of computation, and hence there is no real advantage of one method over another. However, if we need to solve several systems with the same A but different b's, then LU factorization clearly has the advantage over Gaussian elimination (recall that LU factorization is done only once). We illustrate this important fact by the following example.

Example 7.8. For the matrix A considered in Example 4.2, we shall use the LU factorization method to compute its inverse. For this, we need to solve the systems $Ax = e^1$, $Ax = e^2$, $Ax = e^3$, which will provide, respectively, the first, second, and third column of A^{-1}. This in view of Example 7.3 is equivalent to solving the systems

$$
\begin{array}{llll}
y_1 = 1, & y_1 = 1, & 2x_1 - x_2 + x_3 = 1, & x_1 = \frac{11}{28} \\
\frac{3}{2}y_1 + y_2 = 0, & y_2 = -\frac{3}{2}, & \frac{7}{2}x_2 - \frac{13}{2}x_3 = -\frac{3}{2}, & x_2 = \frac{1}{28} \\
\frac{1}{2}y_1 + y_2 + y_3 = 0, & y_3 = 1, & 4x_3 = 1, & x_3 = \frac{1}{4}
\end{array}
$$

$$
\begin{array}{llll}
y_1 = 0, & y_1 = 0, & 2x_1 - x_2 + x_3 = 0, & x_1 = \frac{1}{28} \\
\frac{3}{2}y_1 + y_2 = 1, & y_2 = 1, & \frac{7}{2}x_2 - \frac{13}{2}x_3 = 1, & x_2 = -\frac{5}{28} \\
\frac{1}{2}y_1 + y_2 + y_3 = 0, & y_3 = -1, & 4x_3 = -1, & x_3 = -\frac{1}{4}
\end{array}
$$

$$
\begin{array}{llll}
y_1 = 0, & y_1 = 0, & 2x_1 - x_2 + x_3 = 0, & x_1 = \frac{3}{28} \\
\frac{3}{2}y_1 + y_2 = 0, & y_2 = 0, & \frac{7}{2}x_2 - \frac{13}{2}x_3 = 0, & x_2 = \frac{13}{28} \\
\frac{1}{2}y_1 + y_2 + y_3 = 1, & y_3 = 1, & 4x_3 = 1, & x_3 = \frac{1}{4}.
\end{array}
$$

Problems

7.1. Find LU factorization of the following matrices

(i) $A^1 = \begin{pmatrix} 2 & 0 & 5 \\ 4 & 3 & 12 \\ -14 & 3 & -32 \end{pmatrix}$, (ii) $A^2 = \begin{pmatrix} 2 & 3 & 0 & 5 & 7 \\ 6 & 14 & 3 & 17 & 22 \\ -6 & 1 & 10 & -11 & -16 \end{pmatrix}$

(iii) $A^3 = \begin{pmatrix} 3 & 2 & 5 \\ 6 & 9 & 12 \\ -6 & 6 & -6 \\ 9 & 6 & 15 \end{pmatrix}$, (iv) $A^4 = \begin{pmatrix} 4 & 3 & 2 & 1 \\ 8 & 9 & 6 & 3 \\ 12 & 15 & 12 & 6 \\ 16 & 21 & 16 & 13 \end{pmatrix}$.

7.2. Use LU factorization to find the determinants of the above matrices A^1 and A^4.

7.3. For the matrix

$$
A^5 = \begin{pmatrix} 8 & 8 & 5 & -11 \\ 12 & 13 & 9 & -20 \\ 4 & 3 & 2 & -4 \\ 16 & 18 & 13 & -30 \end{pmatrix}
$$

show that

(i) $A^5 = \begin{pmatrix} 1 & 0 & 0 & 0 \\ \frac{3}{2} & 1 & 0 & 0 \\ \frac{1}{2} & -1 & 1 & 0 \\ 2 & 2 & 0 & 1 \end{pmatrix} \begin{pmatrix} 8 & 8 & 5 & -11 \\ 0 & 1 & \frac{3}{2} & -\frac{7}{2} \\ 0 & 0 & 1 & -2 \\ 0 & 0 & 0 & -1 \end{pmatrix}$

(ii) $A^5 = \begin{pmatrix} 0 & 1 & 0 & 0 \\ 0 & 0 & 1 & 0 \\ 1 & 0 & 0 & 0 \\ 0 & 0 & 0 & 1 \end{pmatrix} \begin{pmatrix} 1 & 0 & 0 & 0 \\ 2 & 1 & 0 & 0 \\ 3 & 2 & 1 & 0 \\ 4 & 3 & 2 & 1 \end{pmatrix} \begin{pmatrix} 4 & 3 & 2 & -4 \\ 0 & 2 & 1 & -3 \\ 0 & 0 & 1 & -2 \\ 0 & 0 & 0 & -1 \end{pmatrix}$

(iii) $\det(A^5) = -8$.

7.4. For each of the above matrices A^i, $i = 1, 2, \cdots, 5$ use LU factorization to solve the systems $A^i x = b^i$, $i = 1, 2, \cdots, 5$, where $b^1 = (3, 14, -12)^t$, $b^2 = (32, 118, -42)^t$, $b^3 = (12, 36, 0, 36)^t$, $b^4 = (10, 26, 45, 66)^t$, $b^5 = (7, 7, 4, 5)^t$.

Answers or Hints

7.1. (i) $\begin{pmatrix} 1 & 0 & 0 \\ 2 & 1 & 0 \\ -7 & 1 & 1 \end{pmatrix} \begin{pmatrix} 2 & 0 & 5 \\ 0 & 3 & 2 \\ 0 & 0 & 1 \end{pmatrix}$.

(ii) $\begin{pmatrix} 1 & 0 & 0 \\ 3 & 1 & 0 \\ -3 & 2 & 1 \end{pmatrix} \begin{pmatrix} 2 & 3 & 0 & 5 & 7 \\ 0 & 5 & 3 & 2 & 1 \\ 0 & 0 & 4 & 0 & 3 \end{pmatrix}$.

(iii) $\begin{pmatrix} 1 & 0 & 0 & 0 \\ 2 & 1 & 0 & 0 \\ -2 & 2 & 1 & 0 \\ 3 & 0 & 0 & 1 \end{pmatrix} \begin{pmatrix} 3 & 2 & 5 \\ 0 & 5 & 2 \\ 0 & 0 & 0 \\ 0 & 0 & 0 \end{pmatrix}$.

(iv) $\begin{pmatrix} 1 & 0 & 0 & 0 \\ 2 & 1 & 0 & 0 \\ 3 & 2 & 1 & 0 \\ 4 & 3 & 1 & 1 \end{pmatrix} \begin{pmatrix} 4 & 3 & 2 & 1 \\ 0 & 3 & 2 & 1 \\ 0 & 0 & 2 & 1 \\ 0 & 0 & 0 & 5 \end{pmatrix}$.

7.2. (i) 6.
(ii) 120.
7.3. (i) Use LU factorization.
(ii) Use PLU factorization.
(iii) Clear from (i) or (ii).
7.4. (i) $(-1, 2, 1)^t$.
(ii) $\frac{1}{20}(233, 58, 50, 0, 0)^t + \frac{1}{20}(-38, -8, 0, 20, 0)^t c + \frac{1}{8}(-31, 2, -6, 0, 8)^t d$.
(iii) $\frac{1}{5}(12, 12, 0)^t + \frac{1}{5}(-7, -2, 5)^t c$.
(iv) $(1, 1, 1, 1)^t$.
(v) $(1, 2, 1, 2)^t$.

Chapter 8

Linear Dependence and Independence

The concept of linear dependence and independence plays an essential role in linear algebra and as a whole in mathematics. These concepts distinguish between two vectors being essentially the same or different. Further, these terms are prerequisites to the geometrical notion of dimension for vector spaces.

Let (V, F) be a vector space, and $S = \{u^1, \cdots, u^n\} \subset V$ be a finite nonempty set. The set of vectors S is said to be *linearly dependent* if and only if there exist n scalars $c_1, \cdots, c_n \in F$, not all zero, such that

$$c_1 u^1 + \cdots + c_n u^n = 0. \tag{8.1}$$

The set S is said to be *linearly independent* if the only solution of (8.1) is the trivial solution $c_1 = \cdots = c_n = 0$. Notice that (8.1) always holds for $c_1 = \cdots = c_n = 0$, and hence, to prove the linear independence of S it is necessary to show that $c_1 = \cdots = c_n = 0$ is the only set of scalars in F for which (8.1) holds. It is clear that if S is linearly dependent, then there are infinitely many choices of scalars c_i, not all zero, such that (8.1) holds. However, to prove the linear dependence of S it suffices to find one set of scalars c_i, not all zero, for which (8.1) holds.

The vectors $u^1, \cdots, u^n$ are said to be linearly dependent or independent accordingly as the set $S = \{u^1, \cdots, u^n\}$ is linearly dependent or independent. An *infinite set of vectors* is said to be *linearly dependent* if and only if it has a finite subset that is linearly dependent. Thus, an infinite set is *linearly independent* if and only if its every finite subset is linearly independent.

Example 8.1. The set of vectors $S = \{e^1, \cdots, e^n\}$ is a linearly independent set in R^n. Indeed, $c_1 e^1 + \cdots + c_n e^n = 0$ immediately implies that $c_1 = \cdots = c_n = 0$. If an $m \times n$ matrix A is in echelon form, then the set of nonzero rows of A (considered as vectors in R^n) is linearly independent.

Example 8.2. The vectors $(11, 19, 21)^t$, $(3, 6, 7)^t$, $(4, 5, 8)^t \in R^3$ are linearly independent. Indeed, $c_1(11, 19, 21)^t + c_2(3, 6, 7)^t + c_3(4, 5, 8)^t = 0$ leads to the system

$$
\begin{aligned}
11c_1 + 3c_2 + 4c_3 &= 0 \\
19c_1 + 6c_2 + 5c_3 &= 0 \\
21c_1 + 7c_2 + 8c_3 &= 0
\end{aligned}
$$

for which $c_1 = c_2 = c_3 = 0$ is the only solution. Now any vector $(a, b, c)^t \in R^3$ can be written as a linear combination of these vectors, i.e., $(a, b, c)^t = \alpha(11, 19, 21)^t + \beta(3, 6, 7)^t + \gamma(4, 5, 8)^t$, where the unknowns α, β, γ can be obtained by solving the linear system

$$
\begin{aligned}
11\alpha + 3\beta + 4\gamma &= a \\
19\alpha + 6\beta + 5\gamma &= b \\
21\alpha + 7\beta + 8\gamma &= c.
\end{aligned}
$$

Theorem 8.1. In any vector space (V, F) the following hold:

1. any set containing 0 is linearly dependent,

2. the set $\{u\}$ is linearly independent if and only if $u \neq 0$,

3. the set $\{u, v\}$ is linearly dependent if and only if $u = cv$, where c is some scalar,

4. every subset S_0 of a linearly independent set S is linearly independent,

5. if S is a finite set of vectors and some subset S_0 of S is linearly dependent, then S is linearly dependent,

6. the set $S = \{u^1, \cdots, u^n\}$ where each $u^k \neq 0$ is linearly dependent if and only if at least one of the vectors u^j is linearly dependent on the preceding vectors $u^1, \cdots, u^{j-1}$.

Proof. 1. If $S = \{0, u^2, \cdots, u^n\}$, then we can write $1 \times 0 + 0 \times u^2 + \cdots + 0 \times u^n = 0$.

2. If $u = 0$, then in view of 1, the set $\{u\}$ is linearly dependent. Conversely, if $\{u\}$ is linearly dependent, then there exists a nonzero scalar c such that $cu = 0$. But, then $0 = c^{-1}0 = c^{-1}(cu) = u$, i.e., $u = 0$.

3. If $S = \{u, v\}$ is linearly dependent, then there exist scalars c_1, c_2 not both zero such that $c_1 u + c_2 v = 0$. If $c_1 = 0$, then $c_2 v = 0$, and hence $v = 0$, which in view of 1. implies that S is linearly dependent. If c_1 and c_2 both are nonzero, then clearly we have $u = -c_1^{-1}c_2 v$. Conversely, in $u = cv$ if $c = 0$, then we have $u = 0$, and if $c \neq 0$, then $u - cv = 0$.

4. Suppose a linear combination of the vectors S_0 is equal to zero, then the addition of all the terms of the form 0 times the vectors in $S \backslash S_0$ is also zero. This gives a linear combination of the vectors of S, which is zero. But since S is linearly independent, all the coefficients in this linear combination must be zero. Thus S_0 is linearly independent.

5. If S is linearly independent, then from 4. S_0 must be linearly independent.

6. Suppose u^j is linearly dependent on $u^1, \cdots, u^{j-1}$, i.e., $u^j = c_1 u^1 + \cdots +$

$c_{j-1}u^{j-1}$, then we have

$$c_1 u^1 + \cdots + c_{j-1}u^{j-1} + (-1)u^j + 0u^{j+1} + \cdots + 0u^n = 0,$$

which implies that S is linearly dependent. Conversely, suppose that S is linearly dependent. We define $S_j = \{u^1, \cdots, u^j\}$, $j = 1, \cdots, n$. In view of 2. S_1 is linearly independent. Let S_i, $2 \le i \le n$ be the first of the S_j that is linearly dependent. Since S_n is linearly dependent, such an S_i exists. We claim that u^i is linearly dependent on the vectors of the set S_{i-1}. For this, we note that there exist scalars $c_1, \cdots, c_i$ not all zero such that $c_1 u^1 + \cdots + c_i u^i = 0$. We note that c_i cannot be zero, otherwise S_{i-1} will be linearly dependent. Thus it follows that $u^i = c_i^{-1}(-c_1)u^1 + \cdots + c_i^{-1}(-c_{i-1})u^{i-1}$. ∎

Theorem 8.2. Let the set $S = \{u^1, \cdots, u^n\}$ span the vector space (V, F).

1. If u^j is linearly dependent on $U = \{u^1, \cdots, u^{j-1}, u^{j+1}, \cdots, u^n\}$, then the set U also spans (V, F).

2. If at least one vector of S is nonzero, then there exists a subset W of S that is linearly independent and spans (V, F).

Proof. 1. See Problem 1.2(ii).

2. If S is linearly independent there is nothing to prove; otherwise, in view of 1. we can eliminate one vector from the set S so that the remaining set will also span V. We can continue this process of eliminating vectors from the set S until either we get (i) a linearly independent subset W that contains at least two vectors and spans V, or (ii) a subset W that contains only one vector, $u^i \ne 0$, say, which spans V. Clearly, in view of Theorem 8.1(2), the set $W = \{u^i\}$ is linearly independent. ∎

The n vector valued functions $u^1(x), \cdots, u^m(x)$ defined in an interval J are said to be *linearly independent* in J if the relation $c_1 u^1(x) + \cdots + c_m u^m(x) = 0$ for all x in J implies that $c_1 = \cdots = c_m = 0$. Conversely, these functions are said to be *linearly dependent* if there exist constants $c_1, \cdots, c_m$ not all zero such that $c_1 u^1(x) + \cdots + c_m u^m(x) = 0$ for all $x \in J$.

Let n vector valued functions $u^1(x), \cdots, u^m(x)$ be linearly dependent in J and $c_k \ne 0$, then we have

$$u^k(x) = -\frac{c_1}{c_k}u^1(x) - \cdots - \frac{c_{k-1}}{c_k}u^{k-1}(x) - \frac{c_{k+1}}{c_k}u^{k+1}(x) - \cdots - \frac{c_m}{c_k}u^m(x),$$

i.e., $u^k(x)$ (and hence at least one of these functions) can be expressed as a linear combination of the remaining $m - 1$ functions. On the other hand, if one of these functions, say, $u^k(x)$, is a linear combination of the remaining $m - 1$ functions, so that

$$u^k(x) = c_1 u^1(x) + \cdots + c_{k-1}u^{k-1}(x) + c_{k+1}u^{k+1}(x) + \cdots + c_m u^m(x),$$

then obviously these functions are linearly dependent. Hence, if two functions are linearly dependent in J, then each one of these functions is identically equal to a constant times the other function, while if two functions are linearly independent then it is impossible to express either function as a constant times the other. The concept of linear independence allows us to distinguish when the given functions are "essentially" different.

Example 8.3. The functions $1, x, \cdots, x^{m-1}$ are linearly independent in every interval J. For this, $c_1 + c_2 x + \cdots + c_m x^{m-1} \equiv 0$ in J implies that $c_1 = \cdots = c_m = 0$. If any c_k were not zero, then the equation $c_1 + c_2 x + \cdots + c_m x^{m-1} = 0$ could hold for at most $m - 1$ values of x, whereas it must hold for all x in J.

Example 8.4. The functions

$$u^1(x) = \begin{bmatrix} e^x \\ e^x \end{bmatrix}, \quad u^2(x) = \begin{bmatrix} e^{2x} \\ 3e^{2x} \end{bmatrix}$$

are linearly independent in every interval J. Indeed,

$$c_1 \begin{bmatrix} e^x \\ e^x \end{bmatrix} + c_2 \begin{bmatrix} e^{2x} \\ 3e^{2x} \end{bmatrix} = 0$$

implies that $c_1 e^x + c_2 e^{2x} = 0$ and $c_1 e^x + 3c_2 e^{2x} = 0$, which is possible only for $c_1 = c_2 = 0$.

Example 8.5. The functions

$$u^1(x) = \begin{bmatrix} \sin x \\ \cos x \end{bmatrix}, \quad u^2(x) = \begin{bmatrix} 0 \\ 0 \end{bmatrix}$$

are linearly dependent.

For the given n vector valued functions $u^1(x), \cdots, u^n(x)$ the determinant $W(u^1, \cdots, u^n)(x)$ or $W(x)$, when there is no ambiguity, defined by

$$\begin{vmatrix} u_1^1(x) & \cdots & u_1^n(x) \\ u_2^1(x) & \cdots & u_2^n(x) \\ \cdots & & \\ u_n^1(x) & \cdots & u_n^n(x) \end{vmatrix}$$

is called the *Wronskian* of these functions. This determinant is closely related to the question of whether or not $u^1(x), \cdots, u^n(x)$ are linearly independent. In fact, we have the following result.

Theorem 8.3. If the Wronskian $W(x)$ of n vector valued functions

$u^1(x), \cdots, u^n(x)$ is different from zero for at least one point in an interval J, then these functions are linearly independent in J.

Proof. Let $u^1(x), \cdots, u^n(x)$ be linearly dependent in J, then there exist n constants $c_1, \cdots, c_n$ not all zero such that $\sum_{i=1}^{n} c_i u^i(x) = 0$ in J. This is the same as saying the homogeneous system of equations $\sum_{i=1}^{n} u_k^i(x) c_i = 0$, $1 \leq k \leq n$, $x \in J$, has a nontrivial solution. However, from Corollary 5.1 this homogeneous system for each $x \in J$ has a nontrivial solution if and only if $W(x) = 0$. But, $W(x) \neq 0$ for at least one x in J, and, therefore $u^1(x), \cdots, u^n(x)$ cannot be linearly dependent. $\blacksquare$

In general the converse of this theorem is not true. For instance, for

$$u^1(x) = \begin{bmatrix} x \\ 1 \end{bmatrix}, \quad u^2(x) = \begin{bmatrix} x^2 \\ x \end{bmatrix},$$

which are linearly independent in any interval J, $W(u^1, u^2)(x) = 0$ in J. This example also shows that $W(u^1, u^2)(x) \neq 0$ in J is not necessary for the linear independence of $u^1(x)$ and $u^2(x)$ in J, and $W(u^1, u^2)(x) = 0$ in J may not imply that $u^1(x)$ and $u^2(x)$ are linearly dependent in J. Thus, the only conclusion we have is $W(x) \neq 0$ in J implies that $u^1(x), \cdots, u^n(x)$ are linearly independent in J and linear dependence of these functions in J implies that $W(x) = 0$ in J.

Problems

8.1. Find if the given vectors are linearly dependent, and if they are, obtain a relation between them:

(i) $u^1 = (1, 3, 4, 2)^t$, $u^2 = (3, -5, 2, 2)^t$, $u^3 = (2, -1, 3, 2)^t$

(ii) $u^1 = (1, 1, 1, 3)^t$, $u^2 = (1, 2, 3, 4)^t$, $u^3 = (2, 3, 4, 9)^t$

(iii) $u^1 = (1, 2, 4)^t$, $u^2 = (2, -1, 3)^t$, $u^3 = (0, 1, 2)^t$, $u^4 = (-3, 7, 2)^t$.

8.2. Let the set $\{u^1, u^2, u^3\}$ in (V, F) be linearly independent. Show that the sets $\{u^1, u^1 + u^2, u^1 + u^2 + u^3\}$ and $\{u^1 + u^2, u^2 + u^3, u^3 + u^1\}$ are also linearly independent.

8.3. Show that rows (columns) of $A \in M^{n \times n}$ are linearly independent if and only if $\det(A) \neq 0$, i.e., the matrix A is invertible. Thus, the homogeneous system $Ax = 0$ has only the trivial solution.

8.4. Show that the nonzero rows of a matrix in echelon form are linearly independent. Further, columns containing pivots are linearly independent.

8.5. Suppose that $\{u^1, \cdots, u^n\}$ is a linearly independent set of vectors in R^n, and A is an $n \times n$ nonsingular matrix. Show that $\{Au^1, \cdots, Au^n\}$ is linearly independent.

8.6. If the rank of A is $n - p$ $(1 \leq p \leq n)$, then show that the system (5.2) with $m = n$ possesses a solution if and only if

$$Bb = 0, \tag{8.2}$$

where B is a $p \times n$ matrix whose row vectors are linearly independent vectors w^i, $1 \leq i \leq p$, satisfying $w^i A = 0$. Further, in the case when (8.2) holds, any solution of (5.2) can be expressed as

$$u = \sum_{i=1}^{p} c_i u^i + Sb,$$

where c_i, $1 \leq i \leq p$, are arbitrary constants, u^i, $1 \leq i \leq p$, are p linearly independent column vectors satisfying $Au^i = 0$, and S is an $n \times n$ matrix independent of b such that $ASv = v$ for any column vector v satisfying $Bv = 0$. The matrix S is not unique.

8.7. The Wronskian of n functions $y_1(x), \cdots, y_n(x)$ which are $(n-1)$ times differentiable in an interval J, is defined by the determinant

$$W(x) = W(y_1, \cdots, y_n)(x) = \begin{vmatrix} y_1(x) & \cdots & y_n(x) \\ y_1'(x) & \cdots & y_n'(x) \\ \cdots & & \\ y_1^{(n-1)}(x) & \cdots & y_n^{(n-1)}(x) \end{vmatrix}.$$

Show that

(i)   if $W(y_1, \cdots, y_n)(x)$ is different from zero for at least one point in J, then the functions $y_1(x), \cdots, y_n(x)$ are linearly independent in J

(ii)   if the functions $y_1(x), \cdots, y_n(x)$ are linearly dependent in J, then the Wronskian $W(y_1, \cdots, y_n)(x) = 0$ in J

(iii)   the converse of (i) as well as of (ii) is not necessarily true.

8.8. Show that the following sets of functions are linearly dependent in any interval

(i) $\{1, \cos 2x, \sin^2 x\}$

(ii) $\{\cos x, \cos 3x, \cos^3 x\}$.

8.9. Show that the following sets of functions are linearly independent in any interval

(i) $\{e^{\mu x}, e^{\nu x}\}$, $\mu \neq \nu$

(ii) $\{e^{\mu x}, xe^{\mu x}\}$.

8.10. Show that the following sets are linearly independent in the given intervals

(i) $\{\sin x, \sin 2x, \cdots\}$, $[0, \pi]$

(ii) $\{1, \cos x, \cos 2x, \cdots\}$, $[0, \pi]$

(iii) $\{1, \sin x, \cos x, \sin 2x, \cos 2x, \cdots\}$, $[-\pi, \pi]$.

8.11. Let $f(x)$ and $g(x)$ be linearly independent in an interval J. Show that the functions $af(x) + bg(x)$ and $cf(x) + dg(x)$ are also linearly independent in J provided $ad - bc \neq 0$.

8.12. Show that

(i)   the set $\{1, x, x^2, \cdots\}$ is linearly independent in the space $\mathcal{P}$ of all polynomials

(ii)   the set $\{u^1, u^2, \cdots\}$, where u^i is the infinite sequence whose i-th term is 1 and all other terms are zero, is linearly independent in the space $\mathcal{S}$ of all real sequences.

Answers or Hints

8.1. (i) Linearly dependent $u^1 + u^2 - 2u^3 = 0$.
(ii)   Linearly independent
(iii) Linearly dependent $-\frac{9}{5}u^1 + \frac{12}{5}u^2 - u^3 + u^4 = 0$.
8.2. $c_1u^1 + c_2(u^1 + u^2) + c_3(u^1 + u^2 + u^3) = 0$ implies $c_1 + c_2 + c_3 = 0$, $c_2 + c_3 = 0$, $c_3 = 0$.
8.3. If the columns $C_1, \cdots, C_n$ of A are linearly independent, then $x_1C_1 + \cdots + x_nC_n = 0$ implies $x_1 = \cdots = x_n = 0$, i.e., the homogeneous system $Ax = 0$ has only the trivial solution. But then by Corollary 5.1, $\det(A) \neq 0$. Conversely, if $\det(A) \neq 0$, then the homogeneous system $(C_1, \cdots, C_n)(x_1, \cdots, x_n)^t = 0$ has only the trivial solution, and hence $C_1, \cdots, C_n$ are linearly independent.
8.4. Recall that in the system (6.4), $j_1 < \cdots < j_r$. Thus, $c_1(1, \tilde{a}_{1,j_1+1}, \cdots, \tilde{a}_{1n}) + \cdots + c_r(1, \tilde{a}_{r,j_r+1}, \cdots, \tilde{a}_{r,n})$ implies that $c_1 = \cdots = c_r = 0$.
8.5. $c_1Au^1 + \cdots + c_nAu^n = 0$ if and only if $A(c_1u^1 + \cdots + c_nu^n) = 0$. Now since A is nonsingular, it is equivalent to $c_1u^1 + \cdots + c_nu^n = 0$.
8.6. Use Problem 8.3.
8.7. (i) If $y_i(x)$, $1 \leq i \leq n$ are linearly dependent, then there exist nontrivial c_i, $1 \leq i \leq n$ such that $\sum_{i=1}^{n} c_iy_i(x) = 0$ for all $x \in J$. Differentiating this, we obtain $\sum_{i=1}^{n} c_iy_i^{(k)}(x) = 0$, $k = 0, 1, \cdots, n-1$ for all $x \in J$. But this implies $W(x) = 0$.
(ii)   Clear from (i).
(iii) Consider the functions $y_1(x) = x^3$, $y_2(x) = x^2|x|$.
8.8. (i) $\cos 2x = 1 - 2\sin^2 x$
(ii) $\cos 3x = 4\cos^3 x - 3\cos x$.

8.9. Use Problem 8.7.

8.10. (i) Use $\int_0^\pi (\sum_{i=1}^n c_i \sin ix) \sin jx dx = \pi c_j/2$

(ii) Similar to (i)

(iii) Similar to (i).

8.11. Let for all $x \in J$, $\alpha(af(x) + bg(x)) + \beta(cf(x) + dg(x)) = 0$. Then, $(a\alpha + c\beta)f(x) + (b\alpha + d\beta)g(x) = 0$. Now the linear independence of $f(x), g(x)$ implies that $a\alpha + c\beta = 0, b\alpha + d\beta = 0$, which implies $\alpha = \beta = 0$ if $ad - bc \neq 0$.

8.12. (i) For each n, $\sum_{i=0}^n a_i x^i = 0$ has at most n roots. Alternatively, using Problem 8.7(i), we have $W(1, x, x^2, \cdots, x^n) = 1(1!)(2!) \cdots (n!)$.

(ii) $0 = \sum_{i=1}^n c_i u^i = (c_1, c_2, \cdots, c_n)^t$.

Chapter 9

Bases and Dimension

In this chapter, for a given vector space, first we shall define a basis and then describe its dimension in terms of the number of vectors in the basis. Here we will also introduce the concept of direct sum of two subspaces.

The set of vectors $S = \{u^1, \cdots, u^n\}$ in a vector space (V, F) is said to form or constitute a *basis* of V over F if and only if S is linearly independent, and generates the space V. Thus, every vector $v \in V$ can be written as a linear combination of the vectors $u^1, \cdots, u^n$, i.e.,

$$v = c_1 u^1 + \cdots + c_n u^n.$$

This representation is unique. Indeed, if $v = d_1 u^1 + \cdots + d_n u^n$ is also a representation, then $0 = v - v = (c_1 - d_1)u^1 + \cdots + (c_n - d_n)u^n$ immediately implies that $c_1 - d_1 = 0, \cdots, c_n - d_n = 0$.

Example 9.1. The set of vectors $S = \{e^1, \cdots, e^n\}$ is a basis of R^n. Indeed, the set S is linearly independent (see Example 8.1), and an arbitrary vector $u = (u_1, \cdots, u_n) \in R^n$ can be written as $u = \sum_{i=1}^{n} u_i e^i$. The set $\{e^1, e^1 + e^2, \cdots, e^1 + \cdots + e^n\}$ is also a basis of R^n.

Theorem 9.1. Let (V, F) be a vector space, and let $S = \{u^1, \cdots, u^n\}$ be a basis of V. Let $v^1, \cdots, v^m$ be vectors in V and assume that $m > n$. Then, $v^1, \cdots, v^m$ are linearly dependent.

Proof. Since S is a basis, there exist scalars a_{ij}, $i = 1, \cdots, m$, $j = 1, \cdots, n$ in F such that

$$v^1 = a_{11} u^1 + \cdots + a_{1n} u^n$$
$$\cdots$$
$$v^m = a_{m1} u^1 + \cdots + a_{mn} u^n.$$

Let $x_1, \cdots, x_m$ be scalars, then

$$x_1 v^1 + \cdots + x_m v^m$$
$$= (x_1 a_{11} + \cdots + x_m a_{m1})u^1 + \cdots + (x_1 a_{1n} + \cdots + x_m a_{mn})u^n.$$

Now since $m > n$, in view of Corollary 5.2, the homogeneous system

$$x_1 a_{11} + \cdots + x_m a_{m1} = 0$$
$$\cdots$$
$$x_1 a_{1n} + \cdots + x_m a_{mn} = 0$$

has a nontrivial solution. For such a solution $(x_1, \cdots, x_m)$ we have $x_1 v^1 + \cdots + x_m v^m = 0$, i.e., the vectors $v^1, \cdots, v^m$ are linearly dependent. ∎

Corollary 9.1. Let (V, F) be a vector space, and let $S = \{u^1, \cdots, u^n\}$ and $T = \{v^1, \cdots, v^m\}$ be the basis of V. Then, $m = n$.

Proof. In view of Theorem 9.1, $m > n$ is impossible. Now since T is also a basis, $n > m$ is also impossible. ∎

Let (V, F) be a vector space, and let $T = \{v^1, \cdots, v^m\} \subset V$. We say that $S = \{v^{k_1}, \cdots, v^{k_n}\} \subseteq T$ is a *maximal linearly independent subset* of T if S is linearly independent, and if $v^i \in T \backslash S$, then $S \cup \{v^i\}$ is linearly dependent.

Theorem 9.2. Let $T = \{v^1, \cdots, v^m\}$ generate the vector space (V, F), and let $S = \{v^{k_1}, \cdots, v^{k_n}\}$ be a maximal linearly independent subset of T. Then S is a basis of V.

Proof. We need to show that $S = \{v^{k_1}, \cdots, v^{k_n}\}$ generates V. For this, first we shall prove that each $v^i \in T \backslash S$ is a linear combination of $v^{k_1}, \cdots, v^{k_n}$. Since $S \cup \{v^i\}$ is linearly dependent, there exist scalars $c_1, \cdots, c_n, x$ not all zero such that

$$c_1 v^{k_1} + \cdots + c_n v^{k_n} + x v^i = 0.$$

Clearly, $x \neq 0$, otherwise, S will be linearly dependent. Thus, it follows that

$$v^i = \frac{-c_1}{x} v^{k_1} + \cdots + \frac{-c_n}{x} v^{k_n},$$

which confirms that v^i is a linear combination of $v^{k_1}, \cdots, v^{k_n}$. Now let $u \in V$, then there exist scalars $\alpha_1, \cdots, \alpha_m$ such that

$$u = \alpha_1 v^1 + \cdots + \alpha_m v^m.$$

In this relation, we replace each $v^i \in T \backslash S$ by a linear combination of $v^{k_1}, \cdots, v^{k_n}$. The resulting relation on grouping the terms will lead to a linear combination of $v^{k_1}, \cdots, v^{k_n}$ for u. Thus S generates V. ∎

Let (V, F) be a vector space, and let $S = \{u^1, \cdots, u^n\} \subset V$ be linearly independent. We say that S is a *maximal linearly independent set* of V if and only if $u \in V \backslash S$, then $S \cup \{u\}$ is linearly dependent.

Theorem 9.3. Let $S = \{u^1, \cdots, u^n\}$ be a maximal linearly independent set of the vector space (V, F). Then, S is a basis of V.

Proof. The proof is the same as in the first part of Theorem 9.2. ∎

A vector space V over the field F is said to be of *finite dimension* n, or n-*dimensional*, written as $\dim V = n$, if and only if V has a basis consisting of n vectors. If $V = \{0\}$, we say V has dimension 0. If V does not have a finite basis, then V is said to be of *infinite dimension*, or *infinite-dimensional*. In the following results we shall consider only finite dimensional vector spaces.

Example 9.2. A field F is a vector space over itself. Its dimension is one, and the element 1 of F forms a basis of F. The space R^n is n-dimensional.

Theorem 9.4. Let (V, F) be an n-dimensional vector space, and let $S = \{u^1, \cdots, u^n\} \subset V$ be linearly independent. Then, S is a basis of V.

Proof. In view of Theorem 9.1, S is a maximal linearly independent set of V. Now, S being a basis follows from Theorem 9.3. ∎

Corollary 9.2. Let (V, F) be a n–dimensional vector space, and let (W, F) be a subspace also of dimension n. Then, $W = V$.

Proof. If $S = \{u^1, \cdots, u^n\}$ is a basis of W, then it must also be a basis of V. ∎

Corollary 9.3. Let (V, F) be an n-dimensional vector space, and let $u^1, \cdots, u^r$, $r < n$ be linearly independent vectors of V. Then, there are vectors $u^{r+1}, \cdots, u^n$ in V such that $S = \{u^1, \cdots, u^n\}$ is a basis of V.

Proof. Since $r < n$, in view of Corollary 9.1, $\{u^1, \cdots, u^r\}$ cannot form a basis of V. Thus there exists a vector $u^{r+1} \in V$ that cannot lie in the subspace generated by $\{u^1, \cdots, u^r\}$. We claim that $\{u^1, \cdots, u^r, u^{r+1}\}$ is linearly independent, i.e., the relation

$$c_1 u^1 + \cdots + c_r u^r + c_{r+1} u^{r+1} = 0, \quad c_i \in F, \quad i = 1, \cdots, r, r+1$$

implies that $c_1 = \cdots = c_r = c_{r+1} = 0$. For this, since $\{u^1, \cdots, u^r\}$ is linearly independent it suffices to show that $c_{r+1} = 0$. Let $c_{r+1} \neq 0$, then from the above relation we have

$$u^{r+1} = -c_{r+1}^{-1}(c_1 u^1 + \cdots + c_r u^r),$$

which contradicts our assumption that u^{r+1} does not lie in the space generated by $\{u^1, \cdots, u^r\}$. Now let us assume that $u^{r+1}, \cdots, u^s$ have been found so that $\{u^1, \cdots, u^r, u^{r+1}, \cdots, u^s\}$ is linearly independent. Then from Theorem 9.1 it follows that $s \leq n$. If we choose s to be maximal, then we have $s = n$, and Theorem 9.4 ensures that $\{u^1, \cdots, u^n\}$ is a basis of V. ∎

In view of our above discussion the set of vectors $S = \{u^1, \cdots, u^m\}$ in R^n cannot span R^n if $m < n$. Also, if $m > n$ the set S is linearly dependent. Further, if $m \geq n$, the set may or may not span R^n. Combining these remarks with Problem 8.4, we find an effective method to extract linearly independent vectors from a given set of vectors. This combination also suggests the possibility of enlarging a set of linearly independent vectors. We illustrate this in the following two examples.

Example 9.3. Since in view of Problems 6.1(iii),

$$\begin{pmatrix} 0 & 1 & 2 & 0 & 3 & 6 \\ 1 & 2 & 5 & 3 & 4 & 3 \\ 1 & 2 & 5 & 6 & 3 & 4 \end{pmatrix} \sim \begin{pmatrix} 1 & 2 & 5 & 3 & 4 & 3 \\ 0 & 1 & 2 & 0 & 3 & 6 \\ 0 & 0 & 0 & 3 & -1 & 1 \end{pmatrix},$$

the vectors $(0,1,1)^t, (1,2,2)^t, (0,3,6)^t$ are linearly independent and form a basis for R^3.

Example 9.4. The vectors $(0,1,1)^t, (1,2,2)^t$ are linearly independent. To find the third linearly independent vector, we consider the matrix whose columns are $(0,1,1)^t, (1,2,2)^t, e^1, e^2, e^3$, and note that

$$\begin{pmatrix} 0 & 1 & 1 & 0 & 0 \\ 1 & 2 & 0 & 1 & 0 \\ 1 & 2 & 0 & 0 & 1 \end{pmatrix} \sim \begin{pmatrix} 1 & 2 & 0 & 1 & 0 \\ 0 & 1 & 1 & 0 & 0 \\ 0 & 0 & 0 & -1 & 1 \end{pmatrix}.$$

Thus, the required vector is $e^2 = (0,1,0)^t$.

Now let U and W be subspaces of a vector space V. Recall that in Problem 1.4 we have defined the sum of U and W as $U+W = \{z : z = u+w$ where $u \in U, w \in W\}$. For this sum we shall prove the following result.

Theorem 9.5. Let U and W be subspaces of a vector space V. Then,

$$\dim(U + W) = \dim U + \dim W - \dim(U \cap W).$$

Proof. In view of Problem 1.3, $U \cap W$ is a subspace of both U and W. Let $\dim U = m$, $\dim W = n$, and $\dim U \cap W = r$. Let $S_1 = \{v^1, \cdots, v^r\}$ be a basis of $U \cap W$. By Corollary 9.3 we can extend S_1 to a basis of U, say $S_2 = \{v^1, \cdots, v^r, u^1, \cdots, u^{m-r}\}$ and to a basis of W, say, $S_3 = \{v^1, \cdots, v^r, w^1, \cdots, w^{n-r}\}$. Now let $S = S_1 \cup S_2 \cup S_3 = \{v^1, \cdots, v^r, u^1, \cdots, u^{m-r}, w^1, \cdots, w^{n-r}\}$. Clearly, S has exactly $m+n-r$ vectors. Thus it suffices to show that S is a basis of $U + W$. For this, we note that S_2 spans U, and S_3 spans W, and hence $S_2 \cup S_3 = S$ spans $U+W$. To show the linear independence of S, suppose that

$$a_1 v^1 + \cdots + a_r v^r + b_1 u^1 + \cdots + b_{m-r} u^{m-r} + c_1 w^1 + \cdots + c_{n-r} w^{n-r} = 0, \quad (9.1)$$

where a_i, b_j, c_k are scalars. Let

$$u = a_1 v^1 + \cdots + a_r v^r + b_1 u^1 + \cdots + b_{m-r} u^{m-r}. \quad (9.2)$$

Then from (9.1), we have

$$u = -c_1 w^1 - \cdots - c_{n-r} w^{n-r}. \quad (9.3)$$

From (9.2) and (9.3), respectively, it follows that $u \in U$ and $u \in W$. Thus,

$u \in U \cap W$, and hence can be written as $u = d_1 v^1 + \cdots + d_r v^r$, where d_i are scalars. Thus, from (9.3) it follows that

$$d_1 v^1 + \cdots + d_r v^r + c_1 w^1 + \cdots + c_{n-r} w^{n-r} = 0.$$

But since S_3 is a basis of W, the above relation implies that $c_1 = \cdots = c_{n-r} = 0$. Substituting this in (9.1), we get

$$a_1 v^1 + \cdots + a_r v^r + b_1 u^1 + \cdots + b_{m-r} u^{m-r} = 0.$$

Finally, since S_2 is a basis of U, the above relation implies that $a_1 = \cdots = a_r = b_1 = \cdots = b_{m-r} = 0$. ∎

The vector space V is said to be a *direct sum* of its subspaces U and W, denoted as $U \oplus W$, if for every $v \in V$ there exist *unique* vectors $u \in U$ and $w \in W$ such that $v = u + w$.

Example 9.5. Consider $V = R^3$, $U = \{(a_1, a_2, 0)^t : a_1, a_2 \in R\}$, $W = \{(0, a_2, a_3)^t : a_2, a_3 \in R\}$. Then, $R^3 = U + W$ but $R^3 \neq U \oplus W$, since sums are not necessarily unique, e.g., $(1, 3, 1)^t = (1, 1, 0)^t + (0, 2, 1)^t = (1, 2, 0)^t + (0, 1, 1)^t$. However, if we let $U = \{(a_1, a_2, 0)^t : a_1, a_2 \in R\}$, $W = \{(0, 0, a_3)^t : a_3 \in R\}$, then $R^3 = U \oplus W$.

Theorem 9.6. The vector space V is the direct sum of its subspaces U and W, i.e., $V = U \oplus W$, if and only if $V = U + W$, and $U \cap W = \{0\}$.

Proof. Suppose $V = U \oplus W$. Then any $v \in V$ can be uniquely written in the form $v = u + w$, where $u \in U$ and $w \in W$. This in particular implies that $V = U + W$. Now let $v \in U \cap W$, then $v = v + 0$ where $v \in U$, $0 \in W$; also, $v = 0 + v$ where $0 \in U$, $v \in W$. But, this in view of the uniqueness implies that $v = 0 + 0 = 0$ and $U \cap W = \{0\}$.

Conversely, suppose that $V = U + W$ and $U \cap W = \{0\}$. If $v \in V$, then there exist $u \in U$ and $w \in W$ such that $v = u + w$. We need to show that this sum is unique. For this, let $v = u' + w'$ where $u' \in U$ and $w' \in W$ is also such a sum. Then, $u + w = u' + w'$, and hence $u - u' = w - w'$. But $u - u' \in U$ and $w - w' \in W$; therefore in view of $U \cap W = \{0\}$ it follows that $u - u' = 0$ and $w - w' = 0$, i.e., $u = u'$ and $w = w'$. ∎

Corollary 9.4. Let U and W be subspaces of a vector space V. Then, $\dim U \oplus W = \dim U + \dim W$.

Problems

9.1. Find the dimension of the following spaces spanned by all

(i) $m \times n$ matrices, and give a basis for this space

(ii) $n \times n$ diagonal matrices

(iii) $n \times n$ upper triangular matrices

(iv) $n \times n$ symmetric matrices

(v)   polynomials of degree $n - 1$, and give a basis for this space.

(vi) n dimensional real vectors $u = (u_1, \cdots, u_n)$ such that $u_1 + \cdots + u_n = 0$

(vii) $n \times n$ real matrices $A = (a_{ij})$ such that $\operatorname{tr}(A) = 0$.

9.2. Suppose that $\{u^1, \cdots, u^n\}$ is a basis for R^n, and A is an $n \times n$ nonsingular matrix. Show that $\{Au^1, \cdots, Au^n\}$ is also a basis for R^n.

9.3. Let (V, F) be a n–dimensional vector space, and let $W \neq \{0\}$ be a subspace. Show that (W, F) has a basis, and its dimension is $\leq n$.

9.4. Let U and W be subspaces of a vector space V, and let $S = \{u^1, \cdots, u^\ell\}$ span U and $T = \{w^1, \cdots, w^k\}$ span W. Show that $S \cup T$ spans $U + W$.

9.5. Let U and W be subspaces of a vector space V, and let $V = U \oplus W$. If $S = \{u^1, \cdots, u^\ell\}$ and $T = \{w^1, \cdots, w^k\}$ are linearly independent subsets of U and W respectively, show that

(i) $S \cup T$ is linearly independent in V

(ii)   if S and T are bases of U and W, then $S \cup T$ is a basis of V.

9.6. Show that the space R over the field of rational numbers Q with the usual operations is infinite dimensional.

9.7. Find $\lambda \in R$ so that the vectors $u^1 = (1, \lambda, 1)^t$, $u^2 = (1, 1, \lambda)^t$, $u^3 = (1, -1, 1)^t$ form a basis of R^3.

9.8. Let $V = \{u \in R^5 : u_1 + 2u_2 - u_3 + u_4 = 0, \ u_1 + u_2 + u_3 + u_5 = 0\}$. Find the dimension of V and find W such that $V \oplus W = R^5$.

Answers or Hints

9.1. (i) mn, $\{E_{ij}\}$ where the $m \times n$ matrix E_{ij} has elements $e_{k\ell} = 1$ if $k = i, \ell = j$ and $e_{kl} = 0$ if $k \neq i, \ell \neq j$.

(ii) n.

(iii) $n(n+1)/2$.

(iv) $n(n+1)/2$.

(v) n, $\{1, x, \cdots, x^{n-1}\}$. This basis is known as *natural*, or *standard basis* for (P_n, R)

(vi) $n - 1$

(vii) $n^2 - 1$.

9.2. $c_1 A u^1 + \cdots + c_n A u^n = 0$ implies $A(c_1 u^1 + \cdots + c_n u^n) = 0$, and since A is nonsingular, $c_1 u^1 + \cdots + c_n u^n = 0$, and hence $c_1 = \cdots = c_n = 0$, i.e., $A u^i$ are linearly independent. Now use Theorem 9.4.

9.3. Let $v^1 \neq 0$ be a vector of W. If $\{v^1\}$ is not a maximal set of linearly independent vectors of W, we can find a vector $v^2 \in W$ such that $\{v^1, v^2\}$ is linearly independent. Continuing in this manner, one vector at a time, we will find an integer $m \leq n$, such that $\{v^1, \cdots, v^m\}$ is a maximal set of linearly independent vectors of W. Now in view of Theorem 9.3 this set is a basis of W.

9.4. If $v \in U + W$, then $v = u + w$, where $u \in U$ and $w \in W$. Clearly, $u = c_1 u^1 + \cdots + c_\ell u^\ell$ and $w = d_1 w^1 + \cdots + d_k w^k$, where c_i, d_j are scalars. Then, $v = c_1 u^1 + \cdots + c_\ell u^\ell + d_1 w^1 + \cdots + d_k w^k$, and hence $S \cup T$ spans $U + W$.

9.5. (i) If $c_1 u^1 + \cdots + c_\ell u^\ell + d_1 w^1 + \cdots + d_k w^k = 0$, then $(c_1 u^1 + \cdots + c_\ell u^\ell) + (d_1 w^1 + \cdots + d_k w^k) = 0 = 0 + 0$, where 0, $c_1 u^1 + \cdots + c_\ell u^\ell \in U$ and 0, $d_1 w^1 + \cdots + d_k w^k \in W$. Now since such a sum for 0 is unique, it follows that $c_1 u^1 + \cdots + c_\ell u^\ell = 0$ and $d_1 w^1 + \cdots + d_k w^k = 0$.

(ii) Use Problem 9.4 and (i).

9.6. The number π cannot be represented by finite rational numbers.

9.7. $\begin{vmatrix} 1 & 1 & 1 \\ \lambda & 1 & -1 \\ 1 & \lambda & 1 \end{vmatrix} = \lambda^2 - 1 \neq 0$, if $\lambda \in R - \{-1, 1\}$.

9.8. For $A = \begin{pmatrix} 1 & 2 & -3 & 1 & 0 \\ 1 & 1 & 1 & 0 & 1 \end{pmatrix}$, we have rank $A = 2$, and hence $\dim V = 2$ with $\{(1, 2, -3, 1, 0), (1, 1, 1, 0, 1)\}$ as a basis. It follows that $\dim W = 3$. Let $w^1 = (0, 1, -1, 2, 1)$. For $A^1 = \begin{pmatrix} 1 & 2 & -3 & 1 & 0 \\ 1 & 1 & 1 & 0 & 1 \\ 0 & 1 & -1 & 2 & 1 \end{pmatrix}$, we have rank $A^1 = 3$, and hence $w^1 \in R^5 - V$. Similarly, $w^2 = (1, -1, 2, 1, 1)$, $w^3 = (0, 1, 0, 1, -1) \in R^5 - V$. Clearly, $W = \text{Span}\{w^1, w^2, w^3\}$.

Chapter 10

Coordinates and Isomorphisms

In this chapter we shall extend the known geometric interpretation of the coordinates of a vector in R^3 to a general vector space. We shall also show how the coordinates of a vector space with respect to one basis can be changed to another basis.

An *ordered basis* of an n-dimensional vector space (V, F) is a fixed sequence of linearly independent vectors that spans V. If $S = \{u^1, \cdots, u^n\}$ is an ordered basis, then as we have noted in Chapter 9, every vector $u \in V$ can be written uniquely as $u = y_1 u^1 + \cdots + y_n u^n$. We call $(y_1, \cdots, y_n)^t \in F^n$ the *coordinates* of u with respect to the basis S, and denote it as $y_S(u)$. Clearly, y_S is a mapping that assigns to each vector $u \in V$ its unique coordinate vector with respect to the ordered basis S. Conversely, each n-tuple $(y_1, \cdots, y_n)^t \in F^n$ corresponds to a unique vector $y_1 u^1 + \cdots + y_n u^n$ in V. Thus the ordered basis S induces a correspondence y_S between V and F^n, which is one-to-one ($u \neq v$ implies $y_S(u) \neq y_S(v)$, equivalently, $y_S(u) \neq y_S(v)$ implies $u \neq v$) and onto (for every $(y_1, \cdots, y_n)^t \in F^n$ there is at least one $u \in V$). It is also easy to observe that

1. $y_S(u + v) = y_S(u) + y_S(v)$ for all $u, \ v \in V$
2. $y_S(\alpha u) = \alpha y_S(u)$ for all scalars α and $u \in V$, and
3. $y_S(u) = 0 \in F^n$ if and only if $u = 0 \in V$.

Hence, the correspondence y_S between V and F^n preserves the vector space operations of vector addition and scalar multiplication. This mapping $y_S : V \to F^n$ is called an *isomorphism* or *coordinate isomorphism*, and the spaces V and F^n are said to be *isomorphic* or *coordinate isomorphic*. One of the major advantages of this concept is that linear dependence and independence, basis, span, and dimension of $U \subseteq V$ can be equivalently discussed for the image set $y_S(U)$.

Example 10.1. In R^n the coordinates of a column vector relative to the ordered basis $\{e^1, \cdots, e^n\}$ are simply the components of the vector. Similarly, the coordinates of a polynomial $a_0 + a_1 x + \cdots + a_{n-1} x^{n-1}$ in $\mathcal{P}_n$ relative to the ordered basis $\{1, x, \cdots, x^{n-1}\}$ are $(a_0, a_1, \cdots, a_{n-1})^t$.

Example 10.2. In view of Example 6.1, we have

$$\frac{13}{28}(2,3,1)^t + \frac{47}{28}(-1,2,3)^t + \frac{21}{28}(1,-5,-2)^t = (0,1,4)^t$$

and hence $(13/28, 47/28, 21/28)^t$ are the coordinates of the vector $(0,1,4)^t$ $\in R^3$ relative to the basis $\{(2,3,1)^t, (-1,2,3)^t, (1,-5,-2)^t\}$.

Now let $S = \{u^1, \cdots, u^n\}$ and $T = \{v^1, \cdots, v^n\}$ be two ordered bases of n-dimensional vector space (V,F). Then, there are unique scalars a_{ij} such that

$$v^j = \sum_{i=1}^{n} a_{ij} u^i, \quad j = 1, \cdots, n. \tag{10.1}$$

Let $(x_1, \cdots, x_n)^t$ be the coordinates of a given vector $u \in V$ with respect to the ordered basis T. Then, we have

$$u = \sum_{j=1}^{n} x_j v^j,$$

and hence from (10.1) it follows that

$$u = \sum_{j=1}^{n} x_j \left(\sum_{i=1}^{n} a_{ij} u^i \right)$$

$$= \sum_{j=1}^{n} \sum_{i=1}^{n} (a_{ij} x_j) u^i$$

$$= \sum_{i=1}^{n} \left(\sum_{j=1}^{n} a_{ij} x_j \right) u^i.$$

Now since the coordinates $(y_1, \cdots, y_n)^t$ of u with respect to the ordered basis S are uniquely determined, we find

$$y_i = \sum_{j=1}^{n} a_{ij} x_j, \quad i = 1, \cdots, n, \tag{10.2}$$

which is exactly the same as (5.1) ((5.2) in matrix notation) with $m = n$. Since S and T are linearly independent sets, $y = (y_1, \cdots, y_n)^t = 0$ if and only if $x = (x_1, \cdots, x_n)^t = 0$. Thus from Corollary 5.1 it follows that the matrix A in (10.2) is invertible, and

$$x = A^{-1} y. \tag{10.3}$$

In terms of coordinate isomorphisms, relations (10.2) and (10.3) for any $u \in V$ can be written as

$$y_S(u) = A x_T(u) \quad \text{and} \quad x_T(u) = A^{-1} y_S(u). \tag{10.4}$$

We summarize the above discussion in the following theorem.

Theorem 10.1. Let $S = \{u^1, \cdots, u^n\}$ and $T = \{v^1, \cdots, v^n\}$ be two ordered bases of an n-dimensional vector space (V, F). Then there exists a unique nonsingular matrix A such that for every vector $u \in V$ relations (10.4) hold.

The matrix A in (10.4) is called the *transition matrix*. A converse of Theorem 10.1 is the following result.

Theorem 10.2. Let $S = \{u^1, \cdots, u^n\}$ be an ordered basis of an n-dimensional vector space (V, F), and let $A \in M^{n \times n}$ be a nonsingular matrix with elements in F. Then there exists an ordered basis $T = \{v^1, \cdots, v^n\}$ of (V, F) such that for every vector $u \in V$ relations (10.4) hold.

Proof. It suffices to show that vectors v^j, $j = 1, \cdots, n$ defined by the equations in (10.1) form a basis of (V, F). For this, let $A^{-1} = B = (b_{ij})$, so that in view of (3.1) – (3.3) and (4.1), we have

$$
\begin{aligned}
\sum_{j=1}^{n} b_{jk} v^j &= \sum_{j=1}^{n} b_{jk} \left(\sum_{i=1}^{n} a_{ij} u^i \right) \\
&= \sum_{j=1}^{n} \left(\sum_{i=1}^{n} a_{ij} b_{jk} \right) u^i \\
&= u^k.
\end{aligned}
$$

Thus, the subspace spanned by the set T contains S, and hence is equal to V. Therefore, T is a basis of (V, F). From the definition of T and Theorem 10.1 it is clear that $y_S(u) = A x_T(u)$ holds. This also implies that $x_T(u) = A^{-1} y_S(u)$. ∎

The equation (10.1) is the same as

$$(v_1^j, \cdots, v_n^j) = a_{1j}(u_1^1, \cdots, u_n^1) + \cdots + a_{nj}(u_1^n, \cdots, u_n^n) \tag{10.5}$$

and hence the j-th column $(a_{1j}, \cdots, a_{nj})^t$ of the matrix A in Theorem 10.1 is the solution of the linear system

$$
\begin{aligned}
u_1^1 a_{1j} + \cdots + u_1^n a_{nj} &= v_1^j \\
\cdots & \\
u_n^1 a_{1j} + \cdots + u_n^n a_{nj} &= v_n^j.
\end{aligned}
$$

Thus the matrix A is the solution of the matrix linear system

$$UA = V, \tag{10.6}$$

where $U = (u_i^j)$, $A = (a_{ij})$, and $V = (v_i^j)$.

Example 10.3. In R^3 consider the ordered bases $S = \{(2,3,1)^t, (-1, 2,3)^t, (1,-5,-2)^t\}$ and $T = \{(1,1,0)^t, (0,1,1)^t, (1,0,1)^t\}$. For these bases the matrix linear system (10.6) is

$$
\begin{pmatrix} 2 & -1 & 1 \\ 3 & 2 & -5 \\ 1 & 3 & -2 \end{pmatrix}
\begin{pmatrix} a_{11} & a_{12} & a_{13} \\ a_{21} & a_{22} & a_{23} \\ a_{31} & a_{32} & a_{33} \end{pmatrix}
=
\begin{pmatrix} 1 & 0 & 1 \\ 1 & 1 & 0 \\ 0 & 1 & 1 \end{pmatrix}.
$$

Thus from Example 6.6 it follows that

$$
\begin{pmatrix} a_{11} & a_{12} & a_{13} \\ a_{21} & a_{22} & a_{23} \\ a_{31} & a_{32} & a_{33} \end{pmatrix}
=
\begin{pmatrix} \frac{11}{28} & \frac{1}{28} & \frac{3}{28} \\ \frac{1}{28} & -\frac{5}{28} & \frac{13}{28} \\ \frac{1}{4} & -\frac{1}{4} & \frac{1}{4} \end{pmatrix}
\begin{pmatrix} 1 & 0 & 1 \\ 1 & 1 & 0 \\ 0 & 1 & 1 \end{pmatrix}
$$

$$
=
\begin{pmatrix} \frac{12}{28} & \frac{4}{28} & \frac{14}{28} \\ -\frac{4}{28} & \frac{8}{28} & \frac{14}{28} \\ 0 & 0 & \frac{14}{28} \end{pmatrix}.
$$

Now consider the vector $(0,1,4)^t \in R^3$ for which the coordinates $(x_1, x_2, x_3)^t$ relative to the basis T are $(-3/2, 5/2, 3/2)^t$. Indeed, we have

$$
-\frac{3}{2}(1,1,0)^t + \frac{5}{2}(0,1,1)^t + \frac{3}{2}(1,0,1)^t = (0,1,4)^t.
$$

The coordinates $(y_1, y_2, y_3)^t$ for $(0,1,4)^t$ relative to the basis S now can be obtained by using (10.4) as follows

$$
\begin{pmatrix} y_1 \\ y_2 \\ y_3 \end{pmatrix}
=
\begin{pmatrix} \frac{12}{28} & \frac{4}{28} & \frac{14}{28} \\ -\frac{4}{28} & \frac{8}{28} & \frac{14}{28} \\ 0 & 0 & \frac{14}{28} \end{pmatrix}
\begin{pmatrix} -\frac{3}{2} \\ \frac{5}{2} \\ \frac{3}{2} \end{pmatrix}
=
\begin{pmatrix} \frac{13}{28} \\ \frac{47}{28} \\ \frac{21}{28} \end{pmatrix},
$$

which are the same as those obtained in Example 10.1.

Example 10.4. In $(\mathcal{P}_3, R)$ consider the ordered bases $S = \{1, x, x^2\}$ and $T = \{1, 1+x, 1+x+x^2\}$. For these bases, (10.5) with $j = 1,2,3$ gives the system

$$
\begin{aligned}
1 &= a_{11}(1) + a_{21}(x) + a_{31}(x^2) = 1(1) + 0(x) + 0(x^2) \\
1+x &= a_{12}(1) + a_{22}(x) + a_{32}(x^2) = 1(1) + 1(x) + 0(x^2) \\
1+x+x^2 &= a_{13}(1) + a_{23}(x) + a_{33}(x^2) = 1(1) + 1(x) + 1(x^2).
\end{aligned}
$$

Thus the matrix A and its inverse A^{-1} in (10.4) for these ordered bases are

$$
A = \begin{pmatrix} 1 & 1 & 1 \\ 0 & 1 & 1 \\ 0 & 0 & 1 \end{pmatrix}, \qquad
A^{-1} = \begin{pmatrix} 1 & -1 & 0 \\ 0 & 1 & -1 \\ 0 & 0 & 1 \end{pmatrix}.
$$

In particular, consider the polynomial $P_2(x) = a + bx + cx^2$ for which the coordinates $(y_1, y_2, y_3)^t$ relative to the basis S are $(a, b, c)^t$. The coordinates $(x_1, x_2, x_3)^t$ for $(a, b, c)^t$ relative to the basis T now can be obtained by using (10.4) as follows:

$$\begin{pmatrix} x_1 \\ x_2 \\ x_3 \end{pmatrix} = \begin{pmatrix} 1 & -1 & 0 \\ 0 & 1 & -1 \\ 0 & 0 & 1 \end{pmatrix} \begin{pmatrix} a \\ b \\ c \end{pmatrix} = \begin{pmatrix} a - b \\ b - c \\ c \end{pmatrix}.$$

Therefore, the polynomial $P_2(x) = a + bx + cx^2$ with respect to the basis T is $P_2(x) = (a - b) + (b - c)(1 + x) + c(1 + x + x^2)$.

Problems

10.1. Find the coordinates of the vector $(a, b, c)^t \in R^3$ with respect to the following ordered bases:

(i) $\{(1, 1, 5)^t, (1, -1, 3)^t, (7, 2, 1)^t\}$
(ii) $\{(1, -1, 3)^t, (1, 1, 5)^t, (7, 2, 1)^t\}$.

10.2. Find the coordinates of the vector $(a, b, c, d)^t \in R^4$ with respect to the following ordered basis: $\{(1, 0, 0, 4)^t, (0, 0, 0, 3)^t, (0, 0, 2, 5)^t, (5, 4, 0, 0)^t\}$.

10.3. Find the coordinates of the matrix $\begin{pmatrix} a & b \\ c & d \end{pmatrix}$ with respect to the ordered basis $\left\{ \begin{pmatrix} 1 & 0 \\ -1 & 0 \end{pmatrix}, \begin{pmatrix} 0 & 1 \\ 1 & 0 \end{pmatrix}, \begin{pmatrix} 1 & -1 \\ 0 & 0 \end{pmatrix}, \begin{pmatrix} 1 & 0 \\ 0 & -1 \end{pmatrix} \right\}$.

10.4. Find the coordinates of the polynomial $P_3(x) = a + bx + cx^2 + dx^3 \in (P_4, R)$ with respect to the following ordered bases:

(i) $\{1, 1 + x, 1 + x + x^2, 1 + x + x^2 + x^3\}$
(ii) $\{1, (1 - x), (1 - x)^2, (1 - x)^3\}$.

10.5. In R^3 consider the ordered bases $S = \{(1, 0, 1)^t, (-1, 0, 0)^t, (0, 1, 2)^t\}$ and $T = \{(-1, 1, 0)^t, (1, 2, -1)^t, (0, 1, 0)^t\}$. For these bases compute the transition matrix and use it to find the coordinates of the vector $(7, 3, 4)^t$ with respect to each of these bases.

10.6. In the vector space of all complex-valued functions on the real line, consider the ordered bases $S = \{1, e^{ix}, e^{-ix}\}$ and $T = \{1, \cos x, \sin x\}$. For these bases compute the transition matrix and use it to find the coordinates of the function $3 + 5 \cos x + 7 \sin x$ with respect to the basis S.

10.7. Consider the vector space (P_n, R) with the ordered basis $S = \{1, x, \cdots, x^{n-1}\}$. Show that

(i) $T = \{L_1(x), \cdots, L_n(x)\}$, where

$$L_i(x) = \frac{(x - x_1) \cdots (x - x_{i-1})(x - x_{i+1}) \cdots (x - x_n)}{(x_i - x_1) \cdots (x_i - x_{i-1})(x_i - x_{i+1}) \cdots (x_i - x_n)}$$

and $x_1 < \cdots < x_n$ are points in R, is also a basis for $(\mathcal{P}_n, R)$

(ii)   the transition matrix for changing the basis S to T is the Vandermonde matrix given in Example 3.5.

10.8. In Theorem 10.2, let $S = \{(5, 0, 3)^t, (3, 5, 7)^t, (-1, -4, 11)^t\}$ and

$$A = \begin{pmatrix} 11 & 3 & 4 \\ 19 & 6 & 5 \\ 21 & 7 & 8 \end{pmatrix}.$$

Find the ordered basis T.

10.9. In Theorem 10.2, let $T = \{(5, 0, 3)^t, (3, 5, 7)^t, (-1, -4, 11)^t\}$ and A be as in Problem 10.8. Find the ordered basis S.

Answers or Hints

10.1. (i) $\frac{1}{58}(-7a + 20b + 9c, 9a - 34b + 5c, 8a + 2b - 2c)^t$.

(ii) $\frac{1}{58}(9a - 34b + 5c, -7a + 20b + 9c, 8a + 2b - 2c)^t$.

10.2. $(a - \frac{5}{4}b, -\frac{4}{3}a + \frac{5}{3}b - \frac{5}{6}c + \frac{1}{3}d, \frac{1}{2}c, \frac{1}{4}b)^t$.

10.3. $\frac{1}{2}(a + b - c + d, a + b + c + d, a - b + c + d, -2d)^t$.

10.4. (i) $(a - b, b - c, c - d, d)^t$.

(ii) $(a + b + c + d, -b - 2c - 3d, c + 3d, -d)^t$.

10.5. $\begin{pmatrix} -2 & -5 & -2 \\ -1 & -6 & -2 \\ 1 & 2 & 1 \end{pmatrix}$, $(-11, -4, 22)^t_T$, $(-2, -9, 3)^t_S$.

10.6. $\begin{pmatrix} 1 & 0 & 0 \\ 0 & \frac{1}{2} & -\frac{i}{2} \\ 0 & \frac{1}{2} & \frac{i}{2} \end{pmatrix}$, $(3, 5, 7)^t_T$, $(3, \frac{5-7i}{2}, \frac{5+7i}{2})^t_S$.

10.7. (i) See Problem 1.5.

(ii) $x^j = x_1^j L_1(x) + \cdots + x_n^j L_n(x), \quad j = 0, 1, \cdots, n - 1$.

10.8. $T = \{(91, 11, 397)^t, (26, 2, 128)^t, (27, -7, 135)^t\}$.

10.9. $S = \{\frac{1}{30}(-83, -263, -213)^t, \frac{1}{30}(46, 76, -114)^t, \frac{1}{30}(9, 69, 219)^t\}$.

Chapter 11

Rank of a Matrix

The rank of a matrix has been briefly defined in Chapter 5. Here we shall give another equivalent definition of the rank and show how this number is directly attached to the dimension of the solution space of the homogeneous linear system (5.3).

For a given matrix $A \in C^{m \times n}$ its rows (columns) generate a subspace of C^n (C^m), called the *row space* (*column space*) of A, and denoted as $R(A)$ ($C(A)$). It is clear that two row (column) equivalent matrices have the same row (column) space. The *row* (*column*) *rank* of the matrix A is the dimension of the row (column) space of A.

Theorem 11.1. For any matrix $A \in C^{m \times n}$ the row rank is equal to its column rank. (This common rank is called the *rank* of A, and is denoted as $r(A)$.)

Proof. Let $v^1, \cdots, v^m$ be the row vectors of A, where $v^i = (a_{i1}, \cdots, a_{in})$, $i = 1, \cdots, m$. Let r be the row rank of A, i.e., $\dim R(A) = r$. Further, let the set of vectors $\{w^1, \cdots, w^r\}$ form a basis for the row space of A, where $w^i = (b_{i1}, \cdots, b_{in})$, $i = 1, \cdots, r$. It follows that

$$v^i = c_{i1}w^1 + \cdots + c_{ir}w^r, \quad i = 1, \cdots, m \qquad (11.1)$$

where c_{ij} are uniquely determined scalars. Now equating the coefficients in (11.1), we get

$$a_{ij} = c_{i1}b_{1j} + \cdots + c_{ir}b_{rj}, \quad i = 1, \cdots, m, \quad j = 1, \cdots, n,$$

which gives

$$\begin{pmatrix} a_{1j} \\ \vdots \\ a_{mj} \end{pmatrix} = b_{1j} \begin{pmatrix} c_{11} \\ \vdots \\ c_{m1} \end{pmatrix} + \cdots + b_{rj} \begin{pmatrix} c_{1r} \\ \vdots \\ c_{mr} \end{pmatrix}, \quad j = 1, \cdots, n.$$

Hence, every column of A is a linear combination of r vectors. Thus, the dimension of the column space of A is at most r, i.e., $\dim C(A) \le r = \dim R(A)$. In exactly the same way we can show that $\dim R(A) \le \dim C(A)$. Therefore, the row and column ranks of A must be equal. ∎

Corollary 11.1. $r(A) = r(A^t) = r(A^H)$.

Theorem 11.2. Let $A \in C^{m \times n}$ and $B \in C^{n \times p}$. Then, $r(AB) \leq \min\{r(A), r(B)\}$.

Proof. Let $A = (a_{ij})$, $B = (b_{ij})$ and $AB = C = (c_{ij})$. Since $c_{ij} = \sum_{k=1}^{n} a_{ik} b_{kj}$, $i = 1, \cdots, m$, $j = 1, \cdots, p$ it follows that

$$
\begin{pmatrix} c_{1j} \\ \vdots \\ c_{mj} \end{pmatrix} = \begin{pmatrix} a_{11} \\ \vdots \\ a_{m1} \end{pmatrix} b_{1j} + \cdots + \begin{pmatrix} a_{1n} \\ \vdots \\ a_{mn} \end{pmatrix} b_{nj}, \quad j = 1, \cdots, p
$$

and hence the columns of AB are dependent on the columns of A; and similarly, the rows of AB are dependent on the rows of B. Thus, we have $C(AB) \subseteq C(A)$ and $R(AB) \subseteq R(B)$, and this immediately implies that $r(AB) \leq r(A)$ and $r(AB) \leq r(B)$. Combining these, we obtain $r(AB) \leq \min\{r(A), r(B)\}$. ∎

Corollary 11.2. 1. If one matrix in the product AB is nonsingular, then the rank of AB is equal to the rank of the other matrix.

2. If P and Q are nonsingular matrices, then $r(PAQ) = r(A)$.

Proof. 1. Let the matrix A be nonsingular. From Theorem 11.2 it follows that $r(AB) \leq r(B)$, and since $B = A^{-1}(AB)$, $r(B) \leq r(AB)$. Thus, $r(AB) = r(B)$.

2. Since P and Q are nonsingular $r(PAQ) = r(AQ) = r(A)$. ∎

Theorem 11.3. Let $A, B \in C^{m \times n}$. Then, $r(A + B) \leq r(A) + r(B)$.

Proof. Let $A = (a_{ij})$, $B = (b_{ij})$, and let $\{u^1, \cdots, u^p\}$ and $\{v^1, \cdots, v^q\}$ be the bases of $C(A)$ and $C(B)$, respectively. Then, it follows that

$$
\begin{pmatrix} a_{1j} + b_{1j} \\ \vdots \\ a_{mj} + b_{mj} \end{pmatrix} = \begin{pmatrix} a_{1j} \\ \vdots \\ a_{mj} \end{pmatrix} + \begin{pmatrix} b_{1j} \\ \vdots \\ b_{mj} \end{pmatrix}
$$

$$
= \sum_{i=1}^{p} \alpha_{ij} u^i + \sum_{i=1}^{q} \beta_{ij} v^i, \quad j = 1, \cdots, n.
$$

Thus, the j-th column of $A + B$ can be expressed as a linear combination of the $p + q$ vectors $u^1, \cdots, u^p, v^1, \cdots, v^q$. Hence, the column space of $A + B$ is generated by the linear combination of these vectors. Therefore, it follows that

$$
\dim C(A + B) = \dim \left\{ \sum_{i=1}^{p} s_i u^i + \sum_{i=1}^{q} t_i v^i \right\},
$$

and hence $r(A + B) \leq p + q$. Clearly, $r(A + B) = p + q$ if the set of column vectors $\{u^1, \cdots, u^p, v^1, \cdots, v^q\}$ is linearly independent. ∎

Corollary 11.3. $r(A + B) \geq |r(A) - r(B)|$.

Proof. Clearly, $r(A) = r(A + B - B) \leq r(A + B) + r(-B) = r(A + B) + r(B)$, and hence $r(A + B) \geq r(A) - r(B)$. Similarly, we have $r(A + B) \geq r(B) - r(A)$. ∎

Now recall that for a given matrix $A \in C^{m \times n}$ the null space $\mathcal{N}(A) \subseteq C^n$ is a vector space. The dimension of $\mathcal{N}(A)$ is called the *nullity* of A and is denoted by $n(A)$.

Theorem 11.4. If $A \in C^{m \times n}$, then $r(A) + n(A) = n$.

Proof. Let B be a row reduced echelon form of the matrix A. Since the rank of the matrix A is $r(A)$, first $r(A)$ rows of B will be nonzero, i.e., B will have $r(A)$ pivot elements. Thus the linear homogeneous system (5.3) in its equivalent form $Bx = 0$ will have $r(A)$ basic variables, and $n - r(A)$ free variables. Hence, the system (5.3) has a set of $n - r(A)$ linearly independent solutions, i.e., $n(A) = n - r(A)$. ∎

Corollary 11.4. $r(A) + n(A^t) = m$.

Corollary 11.5. If $r(A) = n \leq m$, then n columns of A are linearly independent, and $x = 0$ is the only solution of the homogeneous linear system (5.3).

Example 11.1. The row and column echelon form of the matrix

$$A = \begin{pmatrix} 1 & 2 & 1 & 5 \\ 1 & 2 & -1 & 1 \\ 2 & 4 & -3 & 0 \end{pmatrix}$$

are

$$\begin{pmatrix} 1 & 2 & 1 & 5 \\ 0 & 0 & 1 & 2 \\ 0 & 0 & 0 & 0 \end{pmatrix} \quad \text{and} \quad \begin{pmatrix} 1 & 0 & 0 & 0 \\ 1 & 2 & 0 & 0 \\ 2 & 5 & 0 & 0 \end{pmatrix}.$$

Thus, $r(A) = 2$, the set $\{(1, 2, 1, 5), (0, 0, 1, 2)\}$ generates $R(A)$, and the set $\{(1, 1, 2)^t, (0, 2, 5)^t\}$ generates $C(R)$. From the row echelon form it also follows that the set $\{(-2, 1, 0, 0)^t, (-3, 0, -2, 1)^t\}$ generates $\mathcal{N}(A)$, and $n(A) = 2$.

Example 11.2. For the matrix A in Problem 6.1(i), $n = 4, r(A) = 3, n(A) = 1$. For the matrix A in Problem 6.1(ii), $n = 5, r(A) = 4, n(A) = 1$. For the matrix A in Problem 6.1(iii), $n = 6, r(A) = 3, n(A) = 3$.

Now for a given matrix $A \in M^{m \times n}$ let there exist a matrix $L \in M^{n \times m}$ ($R \in M^{n \times m}$) such that $LA = I$ ($AR = I$). Then, L (R) is called a *left inverse* (*right inverse*) of A.

Example 11.3. Consider the matrices

$$A = \begin{pmatrix} 1 & 0 & 1 \\ 0 & 1 & 0 \end{pmatrix}, \quad B = \begin{pmatrix} 0 & 1 \\ 0 & 1 \\ 1 & -1 \end{pmatrix}.$$

Since $AB = I$, B is a *right inverse* of A, and A is a *left inverse* of B. We also note that the matrix

$$\begin{pmatrix} 1 & 0 \\ 0 & 1 \\ 0 & 0 \end{pmatrix}$$

is also a right inverse of A. Hence right as well as left inverse of a given matrix is not unique.

Theorem 11.5. If x is a solution of the nonhomogeneous system (5.2) and the matrix A has a left inverse L, then $x = Lb$. If A has a right inverse, then (5.2) has a solution $x = Rb$.

Proof. Notice that $x = Ix = (LA)x = L(Ax) = Lb$, and $A(Rb) = (AR)b = Ib = b$. ■

Theorem 11.6. Let $A \in M^{m \times n}$ and $r = r(A)$. Then A has a right inverse R if and only if $r = m$ and $m \leq n$.

Proof. We note that finding a right inverse matrix $R = (r_{ij})$ of order $n \times m$ is equivalent to solving the matrix linear system $AR = I$ for which the augmented matrix is $(A|I)$. Now from Theorem 5.12, we know that $AR = I$ has a solution if and only if $r(A) = r(A|I)$. But, since $r(A|I) = r(I) = m$ the matrix A has a right inverse if and only if $r(A) = m$, and thus $m \leq n$. ■

Theorem 11.7. Let $A \in M^{m \times n}$ and $r = r(A)$. Then A has a left inverse L if and only if $r = n$ and $n \leq m$.

Proof. The proof follows from the fact that $LA = I$ if and only if $A^t L^t = I$. ■

Thus, to find the left inverse of A we can consider the augmented matrix $(A^t|I)$.

Example 11.4. To find the right inverse R of the matrix

$$A = \begin{pmatrix} 1 & 2 & 3 & 4 \\ 2 & 3 & 4 & 5 \\ 0 & 4 & 6 & 9 \end{pmatrix} \tag{11.2}$$

we consider the augmented matrix

$$(A|I) = \begin{pmatrix} 1 & 2 & 3 & 4 & 1 & 0 & 0 \\ 2 & 3 & 4 & 5 & 0 & 1 & 0 \\ 0 & 4 & 6 & 9 & 0 & 0 & 1 \end{pmatrix},$$

which in row canonical form appears as

$$\left(\begin{array}{ccc|ccc} 1 & 0 & 0 & -\frac{1}{2} & 1 & 0 & -\frac{1}{2} \\ 0 & 1 & 0 & 0 & -6 & 3 & 1 \\ 0 & 0 & 1 & \frac{3}{2} & 4 & -2 & -\frac{1}{2} \end{array}\right). \tag{11.3}$$

Let $(x_1, x_2, x_3, x_4)^t$ be the first column of R. Then, from (11.3) it follows that

$$x_1 - \frac{1}{2}x_4 = 1, \quad x_2 = -6, \quad x_3 + \frac{3}{2}x_4 = 4,$$

which gives $(1+\frac{1}{2}a, -6, 4-\frac{3}{2}a, a)^t$. Similarly, the second and the third columns of R are given by $(\frac{1}{2}b, 3, -2-\frac{3}{2}b, b)^t$ and $(-\frac{1}{2}+\frac{1}{2}c, 1, -\frac{1}{2}-\frac{3}{2}c, c)^t$, respectively. Thus, the right inverse R of the matrix A is

$$\left(\begin{array}{ccc} 1+\frac{1}{2}a & \frac{1}{2}b & -\frac{1}{2}+\frac{1}{2}c \\ -6 & 3 & 1 \\ 4-\frac{3}{2}a & -2-\frac{3}{2}b & -\frac{1}{2}-\frac{3}{2}c \\ a & b & c \end{array}\right), \tag{11.4}$$

where $a, b, c \in R$.

Example 11.5. In view of the above considerations, left inverse L of the matrix $B = A^t$, where A is the same as in (11.2), is the matrix given in (11.4).

Problems

11.1. For the following matrices find the rank, a basis for the row space, a basis for the column space, and a basis for the nullspace

(i) $\left(\begin{array}{ccccc} 1 & 2 & 3 & 4 & 7 \\ 3 & 4 & 5 & 6 & 9 \\ 5 & 7 & 9 & 11 & 13 \end{array}\right)$ (ii) $\left(\begin{array}{cccccc} 1 & 3 & 5 & 7 & 2 & 4 \\ 2 & 6 & 3 & 5 & 6 & 9 \\ -1 & -3 & 1 & 3 & 5 & 4 \end{array}\right)$

11.2. Let $A \in R^{m \times n}$. Show that $\mathcal{N}(A^t A) = \mathcal{N}(A)$ and $\mathcal{N}(AA^t) = \mathcal{N}(A^t)$.

11.3. Let $A \in R^{m \times n}$. Show that $r(A) = r(A^t) = n$ if and only if $r(A^t A) = n$, $C(A^t) = C(A^t A)$, and $r(A) = m$ if and only if $r(AA^t) = m$.

11.4. Let $A \in R^{m \times n}$ and $B \in R^{n \times m}$ with $n < m$. Show that the matrix AB is singular.

11.5. Let $A \in R^{m \times n}$. Define $Y = \{y \in R^m : y = Ax$ for at least one $x \in R^n\}$. Show that $Y = C(A)$. The subspace $Y \subseteq R^m$ is called the *range* of the matrix A.

11.6. Let $A \in R^{m \times n}$. Show that A has infinitely many right (left) inverses if $r(A) = m < n$ $(r(A) = n < m)$. When does A have exactly one right (left) inverse?

11.7. Let $A \in R^{m \times n}$. Show that A has a right inverse if and only if the columns of A span R^m.

11.8. Let $A \in R^{m \times n}$. Show that A has a left inverse if and only if A^t has a right inverse.

11.9. Let $A \in R^{m \times n}$. Show that if the right inverse R of A exists, then $R = A^t(AA^t)^{-1}$.

11.10. Let $A \in R^{m \times n}$. Show that if the left inverse L of A exists, then $L = (A^t A)^{-1} A^t$.

11.11. Compute right inverse for the following matrices

(i) $\begin{pmatrix} 1 & 2 & 3 & 7 \\ -1 & 4 & -2 & 5 \\ 3 & 7 & 0 & 3 \end{pmatrix}$ (ii) $\begin{pmatrix} 5 & 3 & 2 & 1 \\ 4 & 2 & 1 & 3 \\ 1 & 5 & 3 & 5 \end{pmatrix}$.

11.12. Compute left inverse for the following matrices

(i) $\begin{pmatrix} 1 & 1 & 1 \\ 2 & 3 & 4 \\ 3 & 4 & 7 \\ 4 & 5 & 6 \end{pmatrix}$ (ii) $\begin{pmatrix} 1 & -3 & 5 \\ 2 & 4 & -2 \\ 3 & 5 & 2 \\ 0 & 3 & 7 \end{pmatrix}$.

Answers or Hints

11.1. (i) $r(A) = 3$, $\{(1, 0, -1, -2, 0), (0, 1, 2, 3, 0), (0, 0, 0, 0, 1)\}$, $\{(1, 0, 0)^t, (0, 1, 0)^t, (0, 0, 1)^t\}$, $\{(1, -2, 1, 0, 0)^t, (2, -3, 0, 1, 0)^t\}$.
(ii) $r(A) = 3$, $\{(1, 3, 5, 7, 2, 4), (0, 0, 7, 9, -2, -1), (0, 0, 0, 16, 61, 62)\}$, $\{(1, 0, 0)^t, (0, 1, 0)^t, (0, 0, 1)^t\}$, $\{(-3, 1, 0, 0, 0, 0)^t, (-20, 0, 83, -61, 16, 0)^t, (-40, 0, 82, -62, 0, 16)^t\}$.
11.2. $x \in \mathcal{N}(A) \Rightarrow Ax = 0 \Rightarrow A^t(Ax) = 0 \Rightarrow x \in \mathcal{N}(A^t A) \Rightarrow \mathcal{N}(A) \subseteq \mathcal{N}(A^t A)$. Next, $x \in \mathcal{N}(A^t A) \Rightarrow A^t Ax = 0 \Rightarrow x^t A^t Ax = 0 \Rightarrow (Ax)^t (Ax) = 0 \Rightarrow Ax = 0 \Rightarrow x \in \mathcal{N}(A) \Rightarrow \mathcal{N}(A) \subseteq \mathcal{N}(A^t A)$. For the second part, change A to A^t.
11.3. $A \in R^{m \times n} \Rightarrow A^t A \in R^{n \times n}$. By Theorem 11.4, $r(A) + n(A) = n$ and $r(A^t A) + n(A^t A) = n$. From Problem 11.2, $n(A) = n(A^t A)$. Hence, $r(A) = n$ if and only if $r(A^t A) = n$. $C(A^t) = C(A^t A)$ is equivalent to $r(A^t) = r(A^t A)$. The third part is similar.
11.4. $AB \in R^{m \times m}$. From Theorem 11.2, $r(AB) \leq n < m$.
11.5. See Problem 5.2.
11.6. If for $A \in R^{m \times n}$, $r(A) = m < n$, then from Theorem 11.6, A has a

right inverse R. Thus from Theorem 5.12, $r(A) = r(A|R) = m < n$. Now use Theorem 5.10.

11.7. From Theorem 11.6, the matrix A has a right inverse if and only if e^j, $j = 1, \cdots, m$ are in $C(A)$.

11.8. There exists L such that $LA = I$ if and only if $A^t L^t = I$, i.e., if and only if $L^t = R$ such that $A^t R = I$.

11.9. Since R exists, from Theorem 11.6 it follows that $r(A) = m$. Now from Problem 11.3, we have $r(AA^t) = m$. Since AA^t is an $m \times m$ matrix, $(AA^t)^{-1}$ exists, and hence $(AA^t)(AA^t)^{-1} = I$.

11.10. Similar to Problem 11.9.

11.11. (i) $\dfrac{1}{55}\begin{pmatrix} -14 + 155a & -21 + 155b & 16 + 155c \\ 6 - 90a & 9 - 90b & 1 - 90c \\ 19 - 120a & 1 - 120b & -6 - 120c \\ 55a & 55b & 55c \end{pmatrix}$.

(ii) $\dfrac{1}{8}\begin{pmatrix} 1 + a & 1 + b & -1 + c \\ -11 - 43a & 13 - 43b & 3 - 43c \\ 18 + 58a & -22 + 58b & -2 + 58c \\ 8a & 8b & 8c \end{pmatrix}$.

11.12. (i) $\dfrac{1}{2}\begin{pmatrix} 5 - 4a & -3 - 2a & 1 & 2a \\ -2 - 4b & 4 - 2b & -2 & 2b \\ -1 - 4c & -1 - 2c & 1 & 2c \end{pmatrix}$.

(ii) $\dfrac{1}{38}\begin{pmatrix} 18 + 44a & 31 + 137a & -14 - 106a & 38a \\ -10 + 44b & -13 + 137b & 12 - 106b & 38b \\ -2 + 44c & -14 + 137c & 10 - 106c & 38c \end{pmatrix}$.

Chapter 12

Linear Mappings

In this chapter we shall extend some earlier results to general linear mappings between two vector spaces. These mappings are of general interest and have wide applicability, particularly because they preserve the additive structure of linear combinations. Further, often it is possible to approximate an arbitrary mapping by a linear mapping, which can be managed rather easily.

Throughout what follows, unless specified, we shall consider the finite dimensional vector spaces V and W over the same field F. A mapping $L : V \to W$ is called a *linear mapping* (also known as *linear transformation*, and *linear operator*) if and only if it satisfies the following two conditions:

1. For every pair of vectors $u, v \in V$, $L(u + v) = L(u) + L(v)$.
2. For any scalar $c \in F$, and vector $u \in V$, $L(cu) = cL(u)$.

Thus the mapping L is linear if it preserves the two basic operations of a vector space, namely, the vector addition and the scalar multiplication. In particular, the second condition implies that $L(0) = 0$. Clearly, the above two conditions can be unified as follows: $L : V \to W$ is a linear mapping if and only if for any pair of scalars $\alpha, \beta \in F$ and any pair of vectors $u, v \in V$, $L(\alpha u + \beta v) = \alpha L(u) + \beta L(v)$. An immediate extension of this condition gives $L(c_1 u^1 + \cdots + c_n u^n) = c_1 L(u^1) + \cdots + c_n L(u^n)$ for all $c_i \in F$ and $u^i \in V$. If $W = V$, then L is said to be a *linear mapping on V*.

Example 12.1. For a given matrix $A \in M^{m \times n}$ the mapping $L_A : F^n \to F^m$ defined by $L_A(u) = Au$ is linear. Indeed, from the properties of matrices, for all $u, v \in F^n$ and $c, d \in F$, we have

$$L_A(cu + dv) = A(cu + dv) = cAu + dAv = cL_A(u) + dL_A(v).$$

In particular, when

$$A = \begin{pmatrix} \cos \phi & -\sin \phi \\ \sin \phi & \cos \phi \end{pmatrix}$$

the mapping $L_A : R^2 \to R^2$ rotates the xy-plane through an angle ϕ.

Example 12.2. Consider V as the infinite dimensional space of all real integrable functions on an interval J. Then, $L = \int_J$ is a linear mapping.

Indeed, for all $\alpha, \beta \in R$ and $u, v \in V$, we have

$$L(\alpha u + \beta v) = \int_J (\alpha u(x) + \beta v(x))dx = \alpha \int_J u(x)dx + \beta \int_J v(x)dx = \alpha L(u) + \beta L(v).$$

Similarly, on the space V of all real differentiable functions on an interval J, the mapping $L = d/dx$ is linear.

Example 12.3. The projection mapping $L : R^3 \to R^2$ defined by $L(x, y, z) = (x, y, 0)^t$ is linear. However, the translation mapping $L : R^3 \to R^3$ defined by $L(x, y, z) = (x + 1, y + 2, z + 3)^t$ is not linear (nonlinear), because $L(0) = L(0, 0, 0) = (1, 2, 3)^t \neq 0$. The mapping $L : R^3 \to R^2$ defined by $L(x, y, z) = (|x|, yz)^t$ is also nonlinear.

The linear mappings *zero* (for every $u \in V$, $L(u) = 0$) and *identity* (for every $u \in V$, $L(u) = u$) are respectively denoted as 0 and I.

Theorem 12.1. Let $\{u^1, \cdots, u^n\}$ be a basis of V. Then, for any vectors $w^1, \cdots, w^n \in W$ there exists a unique linear mapping $L : V \to W$ such that $L(u^i) = w^i$, $i = 1, \cdots, n$.

Proof. Let $u \in V$. Since $\{u^1, \cdots, u^n\}$ is a basis of V, there exist unique scalars $c_1, \cdots, c_n$ such that $u = c_1 u^1 + \cdots + c_n u^n$. We claim that $L(u) = c_1 w^1 + \cdots + c_n w^n$ is the required mapping. For this, first we note that L is well-defined because c_i, $i = 1, \cdots, n$ are unique, and since $u^i = 0u^1 + \cdots + 0u^{i-1} + 1u^i + 0u^{i+1} + \cdots + 0u^n$, we have $L(u^i) = w^i$, $i = 1, \cdots, n$. Next, to show that L is linear, let $v = b_1 u^1 + \cdots + b_n u^n \in V$ and $k \in F$, then since $u + v = (c_1 + b_1)u^1 + \cdots + (c_n + b_n)u^n$, and $ku = (kc_1)u^1 + \cdots + (kc_n)u^n$, we have $L(u+v) = (c_1 + b_1)w^1 + \cdots + (c_n + b_n)w^n$ and $L(ku) = (kc_1)w^1 + \cdots + (kc_n)w^n$, and hence it follows that $L(u + v) = L(u) + L(v)$ and $L(ku) = kL(u)$. Finally, to show the uniqueness of L, let $\tilde{L} : V \to W$ be another linear mapping such that $\tilde{L}(u^i) = w^i$, $i = 1, \cdots, n$. Then, for every $u \in V$ we find $\tilde{L}(u) = \tilde{L}(c_1 u^1 + \cdots + c_n u^n) = c_1 \tilde{L}(u^1) + \cdots + c_n \tilde{L}(u^n) = c_1 w^1 + \cdots + c_n w^n = L(u)$, which implies that $L = \tilde{L}$. ∎

Now we shall extend the definition of isomorphism given in Chapter 10 to general vector spaces. The spaces V and W are said to be *isomorphic*, written as $V \simeq W$, if there exists a one-to-one and onto linear mapping $L : V \to W$. This mapping L is called an *isomorphism* between V and W. It is clear that $\simeq$ is an equivalence relation, i.e., 1. $V \simeq V$. 2. If $U \simeq V$, then $V \simeq U$. 3. If $U \simeq V$ and $V \simeq W$, then $U \simeq W$.

Theorem 12.2. $V \simeq W$ if and only if $\dim V = \dim W$.

Proof. Let V and W be n-dimensional vector spaces. Then, in view of Chapter 10, $V \simeq F^n$ and $W \simeq F^n$. Now $V \simeq W$ follows from the fact that $\simeq$ is an equivalent relation. Conversely, if $V \simeq W$, then there exists an isomorphism $L : V \to W$. Assume that $\dim V = n$, and let $S = \{u^1, \cdots, u^n\}$

be a basis of V. It suffices to show that $T = \{L(u^1), \cdots, L(u^n)\}$ is a basis for W. For this, let $w \in W$, then $w = L(u)$ for some $u = c_1 u^1 + \cdots + c_n u^n \in V$. Thus, we have

$$w = L(u) = L(c_1 u^1 + \cdots + c_n u^n) = c_1 L(u^1) + \cdots + c_n L(u^n),$$

and this means that T spans W. Now suppose that $c_1 L(u^1) + \cdots + c_n L(u^n) = 0$. Then, $L(c_1 u^1 + \cdots + c_n u^n) = 0$. Since the mapping L is one-to-one it follows that $c_1 u^1 + \cdots + c_n u^n = 0$. However, since S is linearly independent, we have $c_1 = \cdots = c_n = 0$. This means that T is also linearly independent. ∎

Example 12.4. Since the spaces $\mathcal{P}_4$ and $R^{2 \times 2}$ are of dimension four, in view of Theorem 12.2, $\mathcal{P}_4 \simeq R^{2 \times 2}$. In fact, an isomorphic mapping $L : \mathcal{P}_4 \to R^{2 \times 2}$ can be defined by mapping a basis of $\mathcal{P}_4$ onto a basis of $R^{2 \times 2}$. For this, let $S = \{1, x, x^2, x^3\}$ be a basis of $\mathcal{P}_4$, and

$$T = \left\{ \begin{pmatrix} 1 & 0 \\ 0 & 0 \end{pmatrix}, \begin{pmatrix} 0 & 1 \\ 0 & 0 \end{pmatrix}, \begin{pmatrix} 0 & 0 \\ 1 & 0 \end{pmatrix}, \begin{pmatrix} 0 & 0 \\ 0 & 1 \end{pmatrix} \right\}$$

be a basis of $R^{2 \times 2}$, and define $L : S \to T$ as follows

$$L(1) = \begin{pmatrix} 1 & 0 \\ 0 & 0 \end{pmatrix}, \quad L(x) = \begin{pmatrix} 0 & 1 \\ 0 & 0 \end{pmatrix},$$

$$L(x^2) = \begin{pmatrix} 0 & 0 \\ 1 & 0 \end{pmatrix}, \quad L(x^3) = \begin{pmatrix} 0 & 0 \\ 0 & 1 \end{pmatrix}.$$

Then, it follows that

$$L(a + bx + cx^2 + dx^3) = \begin{pmatrix} a & b \\ c & d \end{pmatrix},$$

i.e., with this isomorphism the polynomial $a + bx + cx^2 + dx^3$ acts like a matrix.

Once again, let $L : V \to W$ be a linear mapping. The set $R(L) = \{w \in W : L(u) = w \text{ for some } u \in V\}$ is called the *range* of L, and the set $\mathcal{N}(L) = \{u \in V : L(u) = 0\}$ is called *null space*, or the *kernel* of the mapping L. Clearly, $\mathcal{N}(L)$ extends the definition of null space of a matrix A given in Chapter 5. The *rank* and the *nullity* of L are respectively defined by

$$\operatorname{rank}(L) = \dim R(L) \quad \text{and} \quad \operatorname{nullity}(L) = \dim \mathcal{N}(L).$$

The following propositions can be established rather easily:

1. $R(L)$ is a subspace of W.
2. $\mathcal{N}(L)$ is a subspace of V.
3. If $u^1, \cdots, u^n$ span V, then $L(u^1), \cdots, L(u^n)$ span $R(L)$.

The linear mapping $L : V \to W$ is said to be *singular* if and only if there exists $0 \neq u \in V$ such that $L(u) = 0$. Thus, L is *nonsingular* if and only if $\mathcal{N}(L) = \{0\}$.

Theorem 12.3. Let $L : V \to W$ be a linear mapping. Then, L is one-to-one if and only if $\mathcal{N}(L) = \{0\}$, i.e., L is nonsingular.

Proof. Since $L(0) = 0$, it follows that $0 \in \mathcal{N}(L)$. If L is one-to-one, then $0 \in V$ is the only vector with image $0 \in W$. Thus, we have $\mathcal{N}(L) = \{0\}$. Conversely, let $\mathcal{N}(L) = \{0\}$. If $L(u) = L(v)$ for $u, v \in V$, then $L(u - v) = 0$. But, this implies that $u - v = 0$, i.e., $u = v$. Hence, L is one-to-one. ∎

Corollary 12.1. The equation $L(u) = w$ has at most one solution if and only if $\mathcal{N}(L) = \{0\}$.

Corollary 12.2. Let A be an $m \times n$ matrix and define $T : R^n \to R^m$ by $Tx = Ax$. Then, T is one-to-one if and only if the columns of A are linearly independent.

Example 12.5. The mapping $L : R^2 \to R^2$ defined by $L(x,y) = (2x + y, x + 2y)^t$ is nonsingular, i.e., one-to-one. For this, it suffices to show that $\mathcal{N}(L) = \{0\}$, i.e., the only solution of $(2x + y, x + 2y)^t = (0,0)^t$, which is equivalent to the system $2x + y = 0$, $x + 2y = 0$ is the zero solution. However, the mapping $\tilde{L} : R^4 \to R^3$ defined by $\tilde{L}(x, y, z, w) = (x + 3y + 4z + 2w, 3x - 5y + 2z + 2w, 2x - y + 3z + 2w)^t$ is singular. Indeed, $(1, \frac{7}{13}, -\frac{15}{13}, 1)^t \in \mathcal{N}(\tilde{L})$.

Theorem 12.4. Let $L : V \to W$ be a linear mapping. Then, $\dim V = \dim R(L) + \dim \mathcal{N}(L)$.

Proof. Suppose $\dim \mathcal{N}(L) = r$ and $\{u^1, \cdots, u^r\}$ is a basis of $\mathcal{N}(L)$. Further, suppose $\dim R(L) = s$ and $\{w^1, \cdots, w^s\}$ is a basis of $R(L)$. Since L is a linear mapping for each $w^j \in R(L)$ there exists $v^j \in V$ such that $L(v^j) = w^j$. It suffices to show that $S = \{u^1, \cdots, u^r, v^1, \cdots, v^s\}$ is a basis of V. For this, let $u \in V$. Then, since $L(u) \in R(L)$ there exist scalars c_j such that $L(u) = c_1 w^1 + \cdots + c_s w^s$. Set $\tilde{u} = c_1 v^1 + \cdots + c_s v^s - u$. Then, it follows that

$$
\begin{aligned}
L(\tilde{u}) &= L(c_1 v^1 + \cdots + c_s v^s - u) \\
&= c_1 L(v^1) + \cdots + c_s L(v^s) - L(u) \\
&= c_1 w^1 + \cdots + c_s w^s - L(u) = 0.
\end{aligned}
$$

But this implies that $\tilde{u} \in \mathcal{N}(L)$, and hence there exist scalars a_i such that

$$
\tilde{u} = a_1 u^1 + \cdots + a_r u^r = c_1 v^1 + \cdots + c_s v^s - u,
$$

which gives

$$
u = c_1 v^1 + \cdots + c_s v^s - a_1 u^1 + \cdots - a_r u^r
$$

and hence the set S spans the space V. To show that S is linearly independent, let

$$
x_1 u^1 + \cdots + x_r u^r + y_1 v^1 + \cdots + y_s v^s = 0, \tag{12.1}
$$

where $x_i, y_j \in F$. Then, we have

$$
\begin{aligned}
0 = L(0) &= L(x_1 u^1 + \cdots + x_r u^r + y_1 v^1 + \cdots + y_s v^s) \\
&= x_1 L(u^1) + \cdots + x_r L(u^r) + y_1 L(v^1) + \cdots + y_s L(v^s).
\end{aligned}
$$
(12.2)

However, since $u^i \in \mathcal{N}(L)$, $L(u^i) = 0$, and $L(v^j) = w^j$, from (12.2) it follows that $y_1 w^1 + \cdots + y_s w^s = 0$. But, since w^j are linearly independent, each $y_j = 0$. Thus, (12.1) reduces to $x_1 u^1 + \cdots + x_r u^r = 0$. However, since u^i are linearly independent, each $x_i = 0$. ∎

For the linear mappings $L : V \to W$ and $G : V \to W$ the sum mapping $L + G : V \to W$, and the scalar product mapping $cL : V \to W$, $c \in F$ are respectively defined as $(L + G)(u) = L(u) + G(u)$ and $(cL)(u) = cL(u)$. It is clear that the mappings $L + G$ and cL are also linear. The collection of all mappings from V to W, denoted as $\mathrm{Hom}(V, W)$ (where Hom stands for *homomorphism*), with the above operations of addition and scalar multiplication forms a vector space. For this space the zero element is the *zero mapping*, denoted as 0 from V to W.

Theorem 12.5. Suppose that $\dim V = n$ and $\dim W = m$. Then, $\dim \mathrm{Hom}(V, W) = nm$.

Proof. Let $S = \{u^1, \cdots, u^n\}$ and $T = \{w^1, \cdots, w^m\}$ be ordered bases for V and W, respectively. For each pair of integers (i, j), $i = 1, \cdots, m$, $j = 1, \cdots, n$ we define a linear mapping $L^{i,j}$ from V to W as follows:

$$
L^{i,j}(u^s) = \left\{ \begin{array}{ll} 0 & \text{if } s \neq j \\ w^i & \text{if } s = j \end{array} \right\} = \delta_{sj} w^i.
$$

In view of Theorem 12.1 such a unique linear mapping exists. Clearly, the set $\{L^{i,j}\}$ contains exactly nm elements, and hence it suffices to show that it is a basis for $\mathrm{Hom}(V, W)$. For this, let $L \in \mathrm{Hom}(V, W)$ be an arbitrary mapping. Suppose $L(u^k) = v^k$, $k = 1, \cdots, n$. Since each $v^k \in W$ is a linear combination of w^i, there exist scalars a_{ik} such that

$$
v^k = a_{1k} w^1 + \cdots + a_{mk} w^m, \quad k = 1, \cdots, n.
$$

Consider the linear mapping

$$
G = \sum_{i=1}^{m} \sum_{j=1}^{n} a_{ij} L^{i,j}.
$$

Clearly, we have

$$
G(u^\tau) = \sum_{i=1}^{m} \sum_{j=1}^{n} a_{ij} L^{i,j}(u^\tau) = \sum_{i=1}^{m} a_{i\tau} w^i = v^\tau = L(u^\tau), \quad \tau = 1, \cdots, n.
$$

But, this from Theorem 12.1 implies that $G = L$. Hence, L is a linear combination of $L^{i,j}$, and this means that $\{L^{i,j}\}$ generates $\text{Hom}(V, W)$. Finally, we need to show that $\{L^{i,j}\}$ is linearly independent. For this, let b_{ij} be scalars such that

$$\sum_{i=1}^{m}\sum_{j=1}^{n} b_{ij} L^{i,j} = 0.$$

Then, for u^k, $k = 1, \cdots, n$ we have

$$0 = 0(u^k) = \sum_{i=1}^{m}\sum_{j=1}^{n} b_{ij} L^{i,j}(u^k) = \sum_{i=1}^{m} b_{ik} w^i.$$

But since w^i are linearly independent, we find $b_{ik} = 0$, $i = 1, \cdots, m$, $k = 1, \cdots, n$. ∎

Now suppose that V, W, and U are vector spaces over the same field F. Let $L : V \to W$ and $G : W \to U$ be linear mappings. We define the composition mapping $G \circ L : V \to U$ by $(G \circ L)(u) = G(L(u))$. Since for any scalars $c_1, c_2 \in F$ and any vectors $u, v \in V$,

$$
\begin{aligned}
(G \circ L)(c_1 u + c_2 v) &= G(L(c_1 u + c_2 v)) = G(c_1 L(u) + c_2 L(v)) \\
&= c_1 G(L(u)) + c_2 G(L(v)) \\
&= c_1 (G \circ L)(u) + c_2 (G \circ L)(v),
\end{aligned}
$$

the composition mapping $G \circ L$ is linear.

Problems

12.1. Let $M \in R^{n \times n}$ be a fixed matrix. Show that the mapping L defined by $L(A) = AM + MA$ for all $A \in R^{n \times n}$ is linear.

12.2. Let $L : V \to W$ be a nonsingular linear mapping. Show that the image of any linearly independent set is linearly independent. Is the converse true?

12.3. Determine if the following linear mappings are singular or nonsingular.

(i) $L : R^3 \to R^3$ defined by $L(x, y, z) = (2x + y, 2y + z, 2z + x)^t$

(ii) $L : R^4 \to R^4$ defined by $L(x, y, z, w) = (21x + 17y + 7z + 10w, 24x + 22y + 6z + 10w, 6x + 8y + 2z + 3w, 5x + 7y + z + 2w)^t$

(iii) $L : R^4 \to R^3$ defined by $L(x, y, z, w) = (x + 2y + 3z + w, x + 3y + 5z - 2w, 3x + 8y + 13z - 3w)^t$.

12.4. Verify Theorem 12.4 for the mappings given in Problem 12.3.

12.5. Let V, W, and U be vector spaces over the same field F, and let $L : V \to W$, $\tilde{L} : V \to W$, $G : W \to U$, and $\tilde{G} : W \to U$ be linear mappings. Show that

(i) $\quad G \circ (L + \tilde{L}) = G \circ L + G \circ \tilde{L}$

(ii) $\quad (G + \tilde{G}) \circ L = G \circ L + \tilde{G} \circ L$

(iii) $\quad c(G \circ L) = (cG) \circ L = G \circ (cL), \quad c \in F.$

12.6. Let V, W, and U be vector spaces over the same field F, and $L : V \to W$ and $G : W \to U$ be linear mappings. Show that $\mathrm{rank}(G \circ L) \le \min\{\mathrm{rank}(G), \mathrm{rank}(L)\}$.

12.7. Let $L : V \to W$ be a linear mapping, and $\dim V = \dim W$. Show that

(i) if L is one-to-one, then it is onto

(ii) if L is onto, then it is one-to-one.

12.8. Let $L : V \to W$ be a linear mapping. Show that

(i) if L is onto, then $\dim V \ge \dim W$

(ii) if L is one-to-one, then $\dim V \le \dim W$.

12.9. A linear mapping $L : V \to W$ is called *invertible* if and only if there exists a unique mapping $L^{-1} : W \to V$ such that $L \circ L^{-1} = I$ and $L^{-1} \circ L = I$. Show that $L : V \to W$ is invertible if and only if L is one-to-one and onto. Moreover, L^{-1} is a linear mapping, and $(L^{-1})^{-1} = L$.

12.10. Find L^{-1}, if it exists, for mappings L given in Problem 12.3.

12.11. Let L be a linear mapping on V. Show that the following are equivalent:

(i) $\quad \mathcal{N} = \{0\}$

(ii) $\quad L$ is one-to-one

(iii) $\quad L$ is onto

(iv) $\quad L$ is invertible.

12.12. Let V, W, U be vector spaces over the same field F, and let $L : V \to W$ and $G : W \to U$ be isomorphisms. Show that $G \circ L$ is invertible, and $(G \circ L)^{-1} = L^{-1} \circ G^{-1}$.

12.13. Let L and G be linear mappings on V. Give an example to show that $L \circ G \ne G \circ L$.

12.14. The vector space $\mathrm{Hom}(V, V)$ is usually denoted as $A(V)$. Clearly, the identity mapping $I : V \to V$ belongs to $A(V)$. Show that

(i)   if $\dim V = n$, then $\dim A(V) = n^2$

(ii)   if $L, G \in A(V)$, then $LG = G \circ L \in A(V)$

(iii) if $L, G \in A(V)$ and $LG = I$, then L is invertible and $G = L^{-1}$

(iv)   if $L \in A(V)$, then $L^n \in A(V)$, $n = 1, 2, \cdots$

(v)   if $L, G, H \in A(V)$ and $c \in F$, then (a) $L(G + H) = LG + LH$ (b) $(G + H)L = GL + HL$ (c) $c(GL) = (cG)L = G(cL)$ (d) $(LG)H = L(GH)$.

12.15. Let L and G be linear mappings given by

$$L(x) = \begin{pmatrix} ax_1 + bx_2 \\ cx_1 + dx_2 \end{pmatrix} \quad \text{and} \quad G(x) = \begin{pmatrix} \alpha x_1 + \beta x_2 \\ \gamma x_1 + \delta x_2 \end{pmatrix},$$

where $ad - bc = |A| \neq 0$. Find the matrix A such that $L^{-1}(G(x)) = Ax$.

12.16. Let $L : R^2 \to R^2$ be a linear mapping. A *line segment* between two vectors u and v in R^2 is defined by $tu + (1 - t)v$, $0 \leq t \leq 1$. A set S in R^2 is called *convex* if for every pair of vectors in S, the line segment between the vectors is in S. Show that

(i)   the image of a line segment under the map L is another line segment

(ii)   if L is an isomorphism and S is convex, then $L(S)$ is a convex set.

Answers or Hints

12.1. Verify directly.

12.2. Suppose $u^1, \cdots, u^n \in V$ are linearly independent, and $c_1 L(u^1) + \cdots + c_n L(u^n) = 0$. Then, $L(c_1 u^1 + \cdots + c_n u^n) = 0$. But since, $\mathcal{N}(L) = \{0\}$ we have $c_1 u^1 + \cdots + c_n u^n = 0$, which implies $c_1 = \cdots = c_n = 0$.

12.3. (i) Nonsingular.

(ii)   Singular.

(ii)   Singular.

12.4. (i) $\dim V = 3$, $n(L) = 0$, $\dim R(L) = 3$.

(ii) $\dim V = 4$, $n(L) = 1$, $\dim R(L) = 3$.

(iii) $\dim V = 4$, $n(L) = 2$, $\dim R(L) = 2$.

12.5. Verify directly.

12.6. $L(V) \subseteq W$, and hence $G(L(V)) \subseteq G(W)$. Thus, $\text{rank}(G \circ L) = \dim[G(L(V))] \leq \dim[G(W)] = \text{rank}(G)$. We also have $\text{rank}(G \circ L) = \dim[G(L(V))] \leq \dim[L(V)] = \text{rank}(L)$.

12.7. (i) One-to-one implies $\mathcal{N}(L) = \{0\}$. Thus $\dim V = \dim W = \dim R(L)$, and hence L is onto.

(ii)   Onto implies $R(L) = W$. Thus $\dim V = \dim W = \dim R(L)$, and hence $\mathcal{N}(L) = \{0\}$.

12.8. (i) In Theorem 12.4, $\dim R(L) = \dim W$ and $\dim \mathcal{N}(L) \geq 0$.

(ii)   In Theorem 12.4, $\dim N(L) = 0$ and $\dim R(L) \leq \dim W$.

12.9. Let L be invertible. Suppose $L(u) = L(v)$ for $u, v \in V$. Then $L^{-1}(L(u)) = L^{-1}(L(v))$, so $u = v$, i.e., L is one-to-one. If $w \in W$, then $L(L^{-1}(w)) = w$, so if we let $L^{-1}(w) = u$, then $L(u) = w$. Thus L is onto. Conversely, suppose L is one-to-one and onto. If $w \in W$, then since L is onto, $w = L(u)$ for some $u \in V$, and since L is one-to-one, u is unique. We define $G : W \to V$ by $G(w) = u$. Clearly, $L(G(w)) = L(u) = w$, so that $L \circ G = I$. Also, since $G(L(u)) = G(w) = u$ we have $G \circ L = I$. Thus G is an inverse of L. To show its uniqueness let $\tilde{G} : W \to V$ be such that $L \circ \tilde{G} = I$ and $\tilde{G} \circ L = I$, then $L(G(w)) = w = L(\tilde{G}(w))$ for any $w \in W$. But since L is one-to-one, we conclude that $G(w) = \tilde{G}(w)$. Hence, $G = \tilde{G}$. In conclusion, $G = L^{-1}$. To show L^{-1} is linear, let $w^1, w^2 \in W$ where $L(v^1) = w^1$ and $L(v^2) = w^2$ for $v^1, v^2 \in V$. Then since for any scalars $c_1, c_2 \in F$, $L(c_1 v^1 + c_2 v^2) = c_1 L(v^1) + c_2 L(v^2) = c_1 w^1 + c_2 w^2$ it follows that $L^{-1}(c_1 w^1 + c_2 w^2) = c_1 v^1 + c_2 v^2 = c_1 L^{-1}(w^1) + c_2 L^{-1}(w^2)$. Finally, since $L \circ L^{-1} = I$ and $L^{-1} \circ L = I$, and inverses are unique, we have $(L^{-1})^{-1} = l$.

12.10. (i) $\frac{1}{9}(4a - 2b + c, a + 4b - 2c, -2a + b + 4c)^t$.

(ii) Does not exist.

(iii) Does not exist.

12.11. Use Theorem 12.3 and Problems 12.7 and 12.9.

12.12. We have $(G \circ L) \circ (L^{-1} \circ G^{-1}) = G \circ (L \circ L^{-1}) \circ G^{-1} = G \circ G^{-1} = I$, $L : V \to W$, $G : W \to U$, $\dim V = m$, $\dim W = n$, $\dim U = p$, $(L^{-1} \circ G^{-1}) \circ (G \circ L) = L^{-1} \circ (G^{-1} \circ G) \circ L = L^{-1} \circ L = I$.

12.13. Consider $L(x, y) = (x, -y)^t$ and $G(x, y) = (y, x)^t$.

12.14. (i) See Theorem 12.5.

(ii) Verify directly.

(iii) $n = \operatorname{rank}(I) = \operatorname{rank}(LG) \leq \operatorname{rank}(L) \leq n$ and hence $\operatorname{rank}(L) = n$. Now use Problem 12.11. If $LL^{-1} = L^{-1}L = I$, then $G = IG = L^{-1}L(G) = L^{-1}(LG) = L^{-1}I = L^{-1}$.

(iv) Verify directly.

(v) Verify directly.

12.15. $\dfrac{1}{|A|} \begin{pmatrix} d\alpha - b\gamma & d\beta - b\delta \\ -c\alpha + a\gamma & -c\beta + a\delta \end{pmatrix}$.

12.16. Use definitions.

Chapter 13

Matrix Representation

In this chapter we shall establish the connection between linear mappings and matrices. Our discussion, in particular, generalizes Theorems 10.1 and 10.2. We shall also introduce the concept of similar matrices, which plays an important role in later chapters.

As in Chapter 12, unless specified, here we shall also consider the finite dimensional vector spaces V and W over the same field F. Let $S = \{u^1, \cdots, u^n\}$ and $T = \{w^1, \cdots, w^m\}$ be ordered bases for V and W, respectively, and $L : V \to W$ be a linear mapping. Then, for each $j = 1, \cdots, n$ there exist unique scalars a_{ij}, $i = 1, \cdots, m$ such that

$$L(u^j) = a_{1j}w^1 + \cdots + a_{mj}w^m = \sum_{i=1}^{m} a_{ij}w^i. \tag{13.1}$$

Clearly, $(a_{1j}, \cdots, a_{mj})^t \in F^m$ are the coordinates of $L(u^j)$ in the ordered basis T. Thus, (13.1) implies that the linear mapping L is completely determined by the scalars a_{ij}. This unique $m \times n$ matrix $A = (a_{ij})$ is called *the transition matrix of L relative to the bases S and T.* Now let $u = x_1 u^1 + \cdots + x_n u^n$ be a vector in V. Then from (13.1) it follows that

$$
\begin{aligned}
L(u) &= L\left(\sum_{j=1}^{n} x_j u^j\right) = \sum_{j=1}^{n} x_j L(u^j) \\
&= \sum_{j=1}^{n} x_j \left(\sum_{i=1}^{m} a_{ij} w^i\right) = \sum_{i=1}^{m} \left(\sum_{j=1}^{n} a_{ij} x_j\right) w^i.
\end{aligned}
$$

Thus, if

$$w = L(u) = y_1 w^1 + \cdots + y_m w^m, \tag{13.2}$$

then we have

$$y_i = \sum_{j=1}^{n} a_{ij} x_j. \tag{13.3}$$

Hence, if $(x_1, \cdots, x_n)_S^t$ are the coordinates of $u \in V$ and $(y_1, \cdots, y_m)_T^t$ are the coordinates of $w = L(u) \in W$, then

$$y = Ax \quad \text{or} \quad y_T(L(u)) = Ax_S(u). \tag{13.4}$$

Therefore, each linear mapping $L : V \to W$ can be completely characterized by the $m \times n$ transition matrix $A = (a_{ij})$ relative to the bases S and T. Often, we shall denote this matrix as A_{ST}.

Example 13.1. In $(\mathcal{P}_3, R)$ and $(\mathcal{P}_2, R)$ consider the ordered bases $S = \{1, x, x^2\}$ and $T = \{1, x\}$, respectively. Let $L : \mathcal{P}_3 \to \mathcal{P}_2$ be the differential mapping, i.e., $L(a + bx + cx^2) = b + 2cx$. For this mapping, we have

$$
\begin{array}{rclcl}
L(1) & = & 0 & = & 0(1) + 0(x) \\
L(x) & = & 1 & = & 1(1) + 0(x) \\
L(x^2) & = & 2x & = & 0(1) + 2(x),
\end{array}
$$

and hence the transition matrix relative to the given bases S and T is

$$
A = \begin{pmatrix} 0 & 1 & 0 \\ 0 & 0 & 2 \end{pmatrix}.
$$

In particular, for the polynomial $P_3 = P_3(x) = 4 + 3x + 7x^2$ we have $x_S(P_3) = (4, 3, 7)^t$, and hence

$$
y_T(L(P_3)) = \begin{pmatrix} 0 & 1 & 0 \\ 0 & 0 & 2 \end{pmatrix} \begin{pmatrix} 4 \\ 3 \\ 7 \end{pmatrix} = \begin{pmatrix} 3 \\ 14 \end{pmatrix}.
$$

This immediately gives $L(P_3) = 3 + 14x$.

Example 13.2. Let $L : R^3 \to R^2$ be the linear mapping defined by $L(x, y, z) = (x + y - z, x - y + z)^t$, and let $S = \{(1, 1, 1)^t, (1, 1, 0)^t, (1, 0, 0)^t\}$ and $T = \{(1, 2)^t, (2, 1)^t\}$. Since

$$
(\alpha, \beta)^t = \frac{1}{3}(2\beta - \alpha)(1, 2)^t + \frac{1}{3}(2\alpha - \beta)(2, 1)^t \tag{13.5}
$$

we have

$$
\begin{array}{rclcl}
L(u^1) & = & (1, 1)^t & = & \dfrac{1}{3}(1, 2)^t + \dfrac{1}{3}(2, 1)^t \\[2mm]
L(u^2) & = & (2, 0)^t & = & -\dfrac{2}{3}(1, 2)^t + \dfrac{4}{3}(2, 1)^t \\[2mm]
L(u^3) & = & (1, 1)^t & = & \dfrac{1}{3}(1, 2)^t + \dfrac{1}{3}(2, 1)^t
\end{array}
$$

and hence the transition matrix relative to the given bases S and T is

$$
A = \begin{pmatrix} \frac{1}{3} & -\frac{2}{3} & \frac{1}{3} \\ \frac{1}{3} & \frac{4}{3} & \frac{1}{3} \end{pmatrix}.
$$

Now since

$$
u = (a, b, c)^t = c(1, 1, 1)^t + (b - c)(1, 1, 0)^t + (a - b)(1, 0, 0)^t
$$

from (13.4) it follows that

$$y_T(L(u)) = Ax_S(u) = A(c, b-c, a-b)^t = \left(c - b + \frac{1}{3}a, b - c + \frac{1}{3}a\right)^t.$$
(13.6)

In view of (13.5), we also have

$$L(u) = (a+b-c, a-b+c)^t = \left(c-b+\frac{1}{3}a\right)(1,2)^t + \left(b-c+\frac{1}{3}a\right)(2,1)^t,$$

which confirms (13.6).

Conversely, assume that an $m \times n$ matrix A is given. Then from the relation (13.4) we can compute the coordinates $y_T(L(u))$ for the unknown mapping L. But, then (13.2) uniquely determines the mapping $L : V \to W$. It can easily be verified that this mapping is linear. In conclusion, we find that for the two given vector spaces V and W with fixed ordered bases S and T, there exists a one-to-one correspondence between the set of all linear mappings $L : V \to W$ and the set of all matrices $M^{m \times n}$.

Now let $S = \{u^1, \cdots, u^n\}$, $\tilde{S} = \{\tilde{u}^1, \cdots, \tilde{u}^n\}$ be ordered bases for V, and $T = \{w^1, \cdots, w^m\}$, $\tilde{T} = \{\tilde{w}^1, \cdots, \tilde{w}^m\}$ be ordered bases for W, and $L : V \to W$ be a linear mapping. Since L can be characterized by $A = A_{ST} = (a_{ij})$ and $B = A_{\tilde{S}\tilde{T}} = (b_{ij})$, it is natural to know the relation between these matrices. For this, let

$$L(u^j) = \sum_{i=1}^{m} a_{ij}w^i, \quad j = 1, \cdots, n \tag{13.7}$$

$$L(\tilde{u}^j) = \sum_{k=1}^{m} b_{kj}\tilde{w}^k, \quad j = 1, \cdots, n \tag{13.8}$$

$$u^j = \sum_{\mu=1}^{n} p_{\mu j}\tilde{u}^\mu, \quad j = 1, \cdots, n \tag{13.9}$$

$$w^j = \sum_{\nu=1}^{m} q_{\nu j}\tilde{w}^\nu, \quad j = 1, \cdots, m. \tag{13.10}$$

Then, we have

$$L(u^j) = L\left(\sum_{\mu=1}^{n} p_{\mu j}\tilde{u}^\mu\right) = \sum_{\mu=1}^{n} p_{\mu j}L(\tilde{u}^\mu) = \sum_{\mu=1}^{n}\sum_{k=1}^{m} p_{\mu j}b_{k\mu}\tilde{w}^k \tag{13.11}$$

and

$$L(u^j) = \sum_{i=1}^{m} a_{ij}w^i = \sum_{i=1}^{m} a_{ij}\left(\sum_{\nu=1}^{m} q_{\nu i}\tilde{w}^\nu\right) = \sum_{i=1}^{m}\sum_{\nu=1}^{m} a_{ij}q_{\nu i}\tilde{w}^\nu. \tag{13.12}$$

On comparing the coefficients of $\tilde{w}^k$ in (13.11) and (13.12), we get

$$\sum_{s=1}^{m} q_{ks}a_{sj} = \sum_{s=1}^{n} b_{ks}p_{sj}, \quad j = 1,\cdots,n, \quad k = 1,\cdots,m. \qquad (13.13)$$

Thus, if $Q = (q_{ks})_{m\times m}$, $A = (a_{sj})_{n\times n}$, $B = (b_{ks})_{m\times n}$, $P = (p_{sj})_{n\times n}$ are the matrices, then (13.13) is the same as

$$QA = BP,$$

and since P is invertible, we have the required relation

$$B = QAP^{-1}. \qquad (13.14)$$

In the special case when $V = W$, and $S = T = \{u^1,\cdots,u^n\}$, $\tilde{S} = \tilde{T} = \{\tilde{u}^1,\cdots,\tilde{u}^n\}$, we have $P = Q$, and then the relation (13.14) reduces to

$$B = PAP^{-1}, \qquad (13.15)$$

which is the same as

$$A = P^{-1}BP. \qquad (13.16)$$

Example 13.3. In addition to L, S, T given in Example 13.2, let $\tilde{S} = \{(1,0,0)^t, (0,1,0)^t, (0,0,1)^t\}$ and $\tilde{T} = \{(1,0)^t, (0,1)^t\}$. Then, we have

$$A = \begin{pmatrix} \frac{1}{3} & -\frac{2}{3} & \frac{1}{3} \\ \frac{1}{3} & \frac{4}{3} & \frac{1}{3} \end{pmatrix}, \quad B = \begin{pmatrix} 1 & 1 & -1 \\ 1 & -1 & 1 \end{pmatrix}, \quad Q = \begin{pmatrix} 1 & 2 \\ 2 & 1 \end{pmatrix}$$

$$P = \begin{pmatrix} 1 & 1 & 1 \\ 1 & 1 & 0 \\ 1 & 0 & 0 \end{pmatrix}, \quad P^{-1} = \begin{pmatrix} 0 & 0 & 1 \\ 0 & 1 & -1 \\ 1 & -1 & 0 \end{pmatrix}.$$

For these matrices relation (13.14) follows immediately.

Example 13.4. Let $L : R^2 \to R^2$ be the linear mapping defined by $L(x,y) = (2x + y, x + 2y)^t$, and let $S = \{(1,1)^t, (1,0)^t\}$ and $\tilde{S} = \{(0,1)^t, (1,1)^t\}$. Since

$$\begin{aligned}
L(1,1) &= (3,3)^t = 3(1,1)^t + 0(1,0)^t \\
L(1,0) &= (2,1)^t = 1(1,1)^t + 1(1,0)^t \\
L(0,1) &= (1,2)^t = 1(0,1)^t + 1(1,1)^t \\
L(1,1) &= (3,3)^t = 0(0,1)^t + 3(1,1)^t \\
(1,1)^t &= 0(0,1)^t + 1(1,1)^t \\
(1,0)^t &= -1(0,1)^t + 1(1,1)^t,
\end{aligned}$$

we have

$$A = \begin{pmatrix} 3 & 1 \\ 0 & 1 \end{pmatrix}, \quad B = \begin{pmatrix} 1 & 0 \\ 1 & 3 \end{pmatrix}, \quad P = \begin{pmatrix} 0 & -1 \\ 1 & 1 \end{pmatrix}, \quad P^{-1} = \begin{pmatrix} 1 & 1 \\ -1 & 0 \end{pmatrix}.$$

For these matrices relation (13.15) follows immediately.

Finally, let A and B be two square matrices of the same order. If there exists a nonsingular matrix P such that (13.16) holds, then the matrices A and B are called *similar*, and the matrix P is called a *similarity matrix*. Clearly, matrices A and B in Example 13.4 are similar. From our above discussion it is clear that two matrices A and B are similar if and only if A and B represent the same linear mapping $L : V \to V$ with respect to two ordered bases for V. It is clear that similarity of matrices is an equivalence relation, i.e., reflexive, symmetric, and transitive.

Problems

13.1. In $(\mathcal{P}_3, R)$ and $(\mathcal{P}_4, R)$ consider the ordered bases $S = \{1, 1+x, 1+x+x^2\}$ and $T = \{x^3 + x^2, x^2 + x, x + 1, 1\}$, respectively. Let $L : \mathcal{P}_3 \to \mathcal{P}_4$ be the linear mapping defined by $L(a + bx + cx^2) = (a + b) + (b + c)x + (c + a)x^2 + (a + b + c)x^3$. Find the transition matrix.

13.2. In R^3 and R^2 consider the ordered bases $\{(1, 1, 1)^t, (1, 1, 0)^t, (1, 0, 0)^t\}$ and $\{(1, 3)^t, (1, 2)^t\}$, respectively. Let $L : R^3 \to R^2$ be the linear mapping $L(x, y, z) = (2x + z, 3y - z)^t$. Find the transition matrix.

13.3. In R^3 and R^4 consider the ordered bases $S = \{(1, 1, 0)^t, (1, 0, 1)^t, (0, 1, 1)^t\}$ and $T = \{(1, 1, 1, 1)^t, (1, 1, 1, 0)^t, (1, 1, 0, 0)^t, (1, 0, 0, 0)^t\}$, respectively. Let $L : R^3 \to R^4$ be the linear mapping $L(x, y, z) = (2x + y + z, x + 2y + z, x + y + 2z, 2x - y - z)^t$. Find the transition matrix.

13.4. In Problem 13.1, instead of L let the 4×3 matrix

$$A = \begin{pmatrix} 1 & -1 & 0 \\ 0 & 3 & -2 \\ 3 & 5 & 2 \\ 2 & 1 & 1 \end{pmatrix}$$

be given. Find the linear mapping $L : \mathcal{P}_3 \to \mathcal{P}_4$.

13.5. In Problem 13.2, instead of L let the 2×3 matrix

$$A = \begin{pmatrix} 2 & -1 & 3 \\ -3 & 2 & 1 \end{pmatrix}$$

be given. Find the linear mapping $L : R^3 \to R^2$.

13.6. In Problem 13.3, instead of L let the 4×3 matrix

$$A = \begin{pmatrix} -1 & 1 & 1 \\ 5 & 3 & -2 \\ 4 & 5 & 2 \\ 2 & 1 & -1 \end{pmatrix}$$

be given. Find the linear mapping $L : R^3 \to R^4$.

13.7. In addition to L, S, T given in Problem 13.1, let $\tilde{S} = \{1, 1-x, 1-x-x^2\}$ and $\tilde{T} = \{x^3 - x^2, x^2 - x, x - 1, 1\}$. Verify the relation (13.14).

13.8. In addition to L, S, T given in Problem 13.2, let $\tilde{S} = \{(2,1,1)^t, (2,1,0)^t, (2,0,0)^t\}$ and $\tilde{T} = \{(3,1)^t, (2,1)^t\}$. Verify the relation (13.14).

13.9. In addition to L, S, T given in Problem 13.3, let $\tilde{S} = \{(0,1,1)^t, (1,0,1)^t, (1,1,0)^t\}$ and $\tilde{T} = \{(1,1,0,0)^t, (0,1,1,0)^t, (0,0,1,1)^t, (0,0,0,1)^t\}$. Verify the relation (13.14).

13.10. Let A and B be similar matrices. Show that

(i) $\det A = \det B$

(ii) $\mathrm{tr}(A) = \mathrm{tr}(B)$

(iii) $\mathrm{rank}(A) = \mathrm{rank}(B)$

(iv) $\mathcal{N}(A) = \mathcal{N}(B)$

(v) AB and BA are similar, provided A or B is nonsingular.

Answers or Hints

13.1. $\begin{pmatrix} 1 & 2 & 3 \\ 0 & -1 & -1 \\ 0 & 2 & 3 \\ 1 & 0 & -1 \end{pmatrix}$.

13.2. $\begin{pmatrix} -4 & -1 & -4 \\ 7 & 3 & 6 \end{pmatrix}$.

13.3. $\begin{pmatrix} 1 & 1 & -2 \\ 1 & 2 & 5 \\ 1 & -1 & 0 \\ 0 & 1 & -1 \end{pmatrix}$.

13.4. $L(a+bx+cx^2) = (5a+b-3c)+(3a+5b-8c)x+(a+b-4c)x^2+(a-2b+c)x^3$.

13.5. $L(x,y,z) = (4x - 3y - 2z, 11x - 10y - z)^t$.

13.6. $L(x,y,z) = \left(10x, 8x + z, \frac{9}{2}x - \frac{1}{2}y - \frac{1}{2}z, -\frac{1}{2}x - \frac{1}{2}y + \frac{3}{2}z\right)^t$.

13.7. $B = \begin{pmatrix} 1 & 0 & -1 \\ 2 & 1 & -1 \\ 2 & 0 & -3 \\ 3 & 0 & -3 \end{pmatrix}$, $Q = \begin{pmatrix} 1 & 0 & 0 & 0 \\ 2 & 1 & 0 & 0 \\ 2 & 2 & 1 & 0 \\ 2 & 2 & 2 & 1 \end{pmatrix}$

$A = \begin{pmatrix} 1 & 2 & 3 \\ 0 & -1 & -1 \\ 0 & 2 & 3 \\ 1 & 0 & -1 \end{pmatrix}$, $P = \begin{pmatrix} 1 & 2 & 2 \\ 0 & -1 & 0 \\ 0 & 0 & -1 \end{pmatrix}$.

13.8. $B = \begin{pmatrix} 1 & -2 & 4 \\ 1 & 5 & -4 \end{pmatrix}$, $Q = \begin{pmatrix} -5 & -3 \\ 8 & 5 \end{pmatrix}$, $A = \begin{pmatrix} -4 & -1 & -4 \\ 7 & 3 & 6 \end{pmatrix}$

$P = \begin{pmatrix} 1 & 0 & 0 \\ 0 & 1 & 0 \\ -\frac{1}{2} & -\frac{1}{2} & \frac{1}{2} \end{pmatrix}$.

13.9. $B = \begin{pmatrix} 2 & 3 & 3 \\ 1 & -1 & 0 \\ 2 & 4 & 2 \\ -4 & -3 & -1 \end{pmatrix}$, $Q = \begin{pmatrix} 1 & 1 & 1 & 1 \\ 0 & 0 & 0 & -1 \\ 1 & 1 & 0 & 1 \\ 0 & -1 & 0 & -1 \end{pmatrix}$

$A = \begin{pmatrix} 1 & 1 & -2 \\ 1 & 2 & 5 \\ 1 & -1 & 0 \\ 0 & 1 & -1 \end{pmatrix}$, $P = \begin{pmatrix} 0 & 0 & 1 \\ 0 & 1 & 0 \\ 1 & 0 & 0 \end{pmatrix}$.

13.10. (i) $\det A = \det P^{-1} \det B \det P = \det B$.

(ii) See Problem 4.6.

(iii) Use Corollary 11.2.

(iv) Follows from Part (iii).

(v) $AB = B^{-1}(BA)B$.

Chapter 14

Inner Products and Orthogonality

In this chapter we shall extend the familiar concept-inner product of two or three dimensional vectors to general vector spaces. Our definition of inner products leads to the generalization of the notion of perpendicular vectors, called orthogonal vectors. We shall also discuss the well-known Gram–Schmidt orthogonalization process.

An *inner product* on (V, C) is a function that assigns to each pair of vectors $u, v \in V$ a complex number, denoted as (u, v), or simply by $u \cdot v$, which satisfies the following axioms:

1. *Positive definite property:* $(u, u) > 0$ if $u \neq 0$, and $(u, u) = 0$ if and only if $u = 0$.

2. *Conjugate symmetric property:* $(u, v) = \overline{(v, u)}$.

3. *Linear property:* $(c_1 u + c_2 v, w) = c_1(u, w) + c_2(v, w)$ for all $u, v, w \in V$ and $c_1, c_2 \in C$.

The vector space (V, C) with an inner product is called a *complex inner product space*. From 2. we have $(u, u) = \overline{(u, u)}$ and hence (u, u) must be real, and from 2. and 3. it immediately follows that $(w, c_1 u + c_2 v) = \bar{c}_1(w, u) + \bar{c}_2(w, v)$.

The definition of a *real inner product space* (V, R) remains the same as above except now for each pair $u, v \in V$, (u, v) is real, and hence in 2. complex conjugates are omitted. In (V, R) the angle $0 \leq \theta \leq \pi$ between the vectors u, v is defined by the relation

$$\cos \theta = \frac{(u, v)}{(u, u)^{1/2}(v, v)^{1/2}}. \tag{14.1}$$

Further, the projection of u onto the vector v is denoted and defined by

$$\text{proj}(u, v) = \text{proj}_v u = \frac{(u, v)}{(v, v)} v. \tag{14.2}$$

Example 14.1. Let $u = (a_1, \cdots, a_n)^t, v = (b_1, \cdots, b_n)^t \in R^n$. The inner product in R^n is defined as

$$(u, v) = u^t v = a_1 b_1 + \cdots + a_n b_n = \sum_{i=1}^{n} a_i b_i = v^t u.$$

The inner product in R^n is also called *dot product* and is denoted as $u \cdot v$. The vector space R^n with the above inner product or dot product is simply called an inner product, or dot product, or *Euclidean n-space*.

Thus, for the vectors $u = (2,3,4)^t$, $v = (1,0,7)^t$ in R^3, we find

$$\cos \theta \;=\; \frac{30}{\sqrt{29}\sqrt{50}} \quad \text{and} \quad \text{proj}(u,v) \;=\; \frac{3}{5}(1,0,7)^t.$$

Let $A = (a_{ij}) \in R^{n \times n}$, then it follows that

$$(Au, v) \;=\; (Au)^t v \;=\; u^t A^t v \;=\; u^t(A^t v) \;=\; (u, A^t v),$$

and hence, we have the relation

$$(Au, v) \;=\; (u, A^t v). \tag{14.3}$$

If A is an orthogonal matrix, then from (14.3) it immediately follows that $(Au, Av) = (u, v)$.

Clearly, for $1 \times n$ vectors $u = (a_1, \cdots, a_n), v = (b_1, \cdots, b_n)$ the above definition of inner product is $(u, v) = uv^t = \sum_{i=1}^{n} a_i b_i = vu^t$.

Example 14.2. Let $u = (a_1, \cdots, a_n)^t, v = (b_1, \cdots, b_n)^t \in C^n$. The standard inner (*dot*) product in C^n is defined as

$$(u, v) \;=\; u^t \bar{v} \;=\; a_1 \bar{b}_1 + \cdots + a_n \bar{b}_n \;=\; \sum_{i=1}^{n} a_i \bar{b}_i \;=\; v^H u.$$

The vector space C^n with the above inner product is called a *unitary space*.

Clearly, for $1 \times n$ vectors $u = (a_1, \cdots, a_n), v = (b_1, \cdots, b_n)$ the above definition of inner product is $(u, v) = u\bar{v}^t = uv^H = \sum_{i=1}^{n} a_i \bar{b}_i$.

Example 14.3. In the vector space $(C^{m \times n}, C)$ an inner product for each pair of $m \times n$ matrices $A = (a_{ij}), B = (b_{ij})$ is defined as

$$(A, B) \;=\; \text{tr}(B^H A) \;=\; \sum_{i=1}^{m} \sum_{j=1}^{n} \bar{b}_{ij} a_{ij}.$$

Example 14.4. In the vector space of complex-valued continuous functions $C[a, b]$ an inner product for each pair of functions f, g is defined as

$$(f, g) \;=\; \int_a^b f(x) \bar{g}(x) dx.$$

Let V be an inner product space. Two vectors $u, v \in V$ are said to be *orthogonal* if and only if $(u, v) = 0$. For example, if $u, v \in R^n$, then the vector $(u - \mathrm{proj}(u, v))$ is orthogonal to v. Indeed, we have

$$(u - \mathrm{proj}(u, v), v) = (u, v) - \frac{(u, v)}{(v, v)}(v, v) = 0.$$

A subset S of V is said to be *orthogonal* if and only if every pair of vectors in S is orthogonal, i.e., if $u, v \in S$, $u \neq v$ then $(u, v) = 0$. Clearly, $0 \in V$ is orthogonal to every $u \in V$, since $(0, u) = (0u, u) = 0(u, u) = 0$. Conversely, if v is orthogonal to every $u \in V$, then in particular $(v, v) = 0$, and hence $v = 0$.

The subset $\hat{S}$ is called *orthonormal* if $\hat{S}$ is orthogonal and for every $\hat{u} \in \hat{S}$, $(\hat{u}, \hat{u}) = 1$. If S is an orthogonal set and $u \in S$, then the set $\hat{S}$ of vectors $\hat{u} = u/(u, u)^{1/2}$ is orthonormal. Indeed, if $u, v \in S$, then for $\hat{u} = u/(u, u)^{1/2}$, $\hat{v} = v/(v, v)^{1/2} \in \hat{S}$, we have

$$(\hat{u}, \hat{v}) = \left(\frac{u}{(u, u)^{1/2}}, \frac{v}{(v, v)^{1/2}} \right) = \frac{1}{(u, u)^{1/2}(v, v)^{1/2}}(u, v) = 0$$

and

$$(\hat{u}, \hat{u}) = \left(\frac{u}{(u, u)^{1/2}}, \frac{u}{(u, u)^{1/2}} \right) = \frac{1}{(u, u)^{1/2}(u, u)^{1/2}}(u, u) = 1.$$

The above process of normalizing the vectors of an orthogonal set is called *orthonormalization*.

Example 14.5. From (14.1) it is clear that in the inner product space R^n two vectors $u = (a_1, \cdots, a_n)^t, v = (b_1, \cdots, n_n)^t$ are orthogonal if and only if $(u, v) = \sum_{i=1}^{n} a_i b_i = 0$. The subset $S = \{u, v, w\} = \{(1, 2, 0, -1)^t, (5, 2, 4, 9)^t, (-2, 2, -3, 2)^t\}$ of R^4 is orthogonal. For this, it suffices to note that $u^t v = v^t w = w^t u = 0$. This set can be orthonormalized, to obtain

$$\hat{S} = \left\{ \frac{u}{(u, u)^{1/2}}, \frac{v}{(v, v)^{1/2}}, \frac{w}{(w, w)^{1/2}} \right\} = \left\{ \left(\frac{1}{\sqrt{6}}, \frac{2}{\sqrt{6}}, 0, \frac{-1}{\sqrt{6}} \right)^t, \right.$$
$$\left. \left(\frac{5}{\sqrt{126}}, \frac{2}{\sqrt{126}}, \frac{4}{\sqrt{126}}, \frac{9}{\sqrt{126}} \right)^t, \left(\frac{-2}{\sqrt{21}}, \frac{2}{\sqrt{21}}, \frac{-3}{\sqrt{21}}, \frac{2}{\sqrt{21}} \right)^t \right\}.$$

The set $\{e^1, \cdots, e^n\}$ is orthonormal.

Let S be a subset of an inner product space V. The *orthogonal complement* of S, denoted as $S^\perp$ (read as "S perp") consists of those vectors in V that are orthogonal to every vector $v \in S$, i.e., $S^\perp = \{u \in V : (u, v) = 0 \text{ for every } v \in S\}$. In particular, for a given vector $v \in V$, we have $v^\perp = \{u \in V : (u, v) = 0\}$, i.e., $v^\perp$ consists of all those vectors of V that are orthogonal to v.

For a given subset S of an inner product space V it is clear that $0 \in S^\perp$, as 0 is orthogonal to every vector in V. Further, if $u, w \in S^{\perp}$, then for all scalars α, β and $u \in S$, we have $(\alpha v + \beta w, u) = \alpha(v, u) + \beta(w, u) = 0$, i.e., $\alpha v + \beta w \in S^{\perp}$. Thus, $S^{\perp}$ is a subspace of V.

Example 14.6. Extending the geometric definition of a plane in R^3, an equation of the form $a_1 x_1 + \cdots + a_n x_n = c$ is called a *hyperplane* in R^n. If $c = 0$, then the hyperplane *passes through the origin*. Let $a^i = (a_{i1}, \cdots, a_{in})^t \in R^n$, $i = 1, \cdots, m$ and $x = (x_1, \cdots, x_n)^t \in R^n$. Then, the homogeneous system (5.3) can be written as $((a^i)^t, x) = 0$, $i = 1, \cdots, m$. Thus, geometrically, the solution space $\mathcal{N}(A)$ of the system (5.3) consists of all vectors x in R^n that are orthogonal to every row vector of A, i.e., $\mathcal{N}(A)$ is the orthogonal complement of $R(A)$.

Theorem 14.1. Let (V, C) be an inner product space, and let $S = \{u^1, \cdots, u^n\}$ be an orthogonal subset of nonzero vectors. Then, S is linearly independent.

Proof. Suppose

$$c_1 u^1 + \cdots + c_{i-1} u^{i-1} + c_i u^i + c_{i+1} u^{i+1} + \cdots + c_n u^n = 0. \qquad (14.4)$$

Taking the inner product of (14.4) with u^i, $i = 1, \cdots, n$ we get

$$
\begin{aligned}
0 = (0, u^i) &= (c_1 u^1 + \cdots + c_{i-1} u^{i-1} + c_i u^i + c_{i+1} u^{i+1} + \cdots + c_n u^n, u^i) \\
&= \sum_{k=1, k \neq i}^{n} c_k (u^i, u^k) + c_i (u^i, u^i) \\
&= 0 + c_i (u^i, u^i).
\end{aligned}
$$

Hence, $c_i = 0$, $i = 1, \cdots, n$; and therefore, S is linearly independent. ∎

Corollary 14.1. Let (V, C) be an inner product space, and let $S = \{u^1, \cdots, u^n\}$ be an orthogonal subset of nonzero vectors. If S generates V, then S is a basis (*orthogonal basis*) for V.

The importance of orthogonal bases lies on the fact that working with these bases requires minimum computation.

Theorem 14.2. Let $S = \{u^1, \cdots, u^n\}$ be an orthogonal basis for an inner product space (V, C). Then, for any vector $v \in V$,

$$v = \frac{(v, u^1)}{(u^1, u^1)} u^1 + \frac{(v, u^2)}{(u^2, u^2)} u^2 + \cdots + \frac{(v, u^n)}{(u^n, u^n)} u^n = \sum_{i=1}^{n} \operatorname{proj}_{u^i} v. \qquad (14.5)$$

Proof. Suppose $v = c_1 u^1 + \cdots + c_n u^n$. Then, as in Theorem 14.1, it follows that $(v, u^i) = c_i (u^i, u^i)$, $i = 1, \cdots, n$. ∎

The relation (14.5) is called the *Fourier expansion* of v in terms of the orthogonal basis S, and the scalars $c_i = (v, u^i)/(u^i, u^i)$, $i = 1, \cdots, n$ are called the *Fourier coefficients* of v.

Theorem 14.3. Let (V, C) be an inner product space, and let $\{u^1, \cdots, u^r\}$ be an orthogonal subset of nonzero vectors. Let $v \in V$, and define

$$\tilde{v} = v - (c_1 u^1 + \cdots + c_r u^r), \quad c_i = \frac{(v, u^i)}{(u^i, u^i)}.$$

Then, $\tilde{v}$ is orthogonal to $u^1, \cdots, u^r$.

Proof. It suffices to notice that for each $i = 1, \cdots, r$,

$$
\begin{aligned}
(\tilde{v}, u^i) &= (v - (c_1 u^1 + \cdots + c_r u^r), u^i) \\
&= (v, u^i) - c_i(u^i, u^i) = (v, u^i) - \frac{(v, u^i)}{(u^i, u^i)}(u^i, u^i) = 0. \quad \blacksquare
\end{aligned}
$$

Theorem 14.4 (Gram–Schmidt orthogonalization process). Let (V, C) be an inner product space, and let $S = \{u^1, \cdots, u^n\}$ be a basis. Then, $T = \{v^1, \cdots, v^n\}$, where

$$v^i = u^i - (c_{i1}v^1 + \cdots + c_{i,i-1}v^{i-1}), \quad c_{ij} = \frac{(u^i, v^j)}{(v^j, v^j)}, \tag{14.6}$$

$$i = 1, \cdots, n, \quad j = 1, \cdots, i-1$$

is an orthogonal basis.

Proof. The proof follows from Theorem 14.3. $\quad \blacksquare$

Example 14.7. Consider the basis $\{(0, 1, 1)^t, (1, 0, 1)^t, (1, 1, 0)^t\}$ for R^3. From (14.6), we have

$$
\begin{aligned}
v^1 &= u^1 = (0, 1, 1)^t \\
v^2 &= u^2 - \frac{(u^2, v^1)}{(v^1, v^1)}v^1 = (1, 0, 1)^t - \frac{1}{2}(0, 1, 1)^t = \left(1, -\frac{1}{2}, \frac{1}{2}\right)^t \\
v^3 &= u^3 - \frac{(u^3, v^1)}{(v^1, v^1)}v^1 - \frac{(u^3, v^2)}{(v^2, v^2)}v^2 \\
&= (1, 1, 0)^t - \frac{1}{2}(0, 1, 1)^t - \frac{1}{3}\left(1, -\frac{1}{2}, \frac{1}{2}\right)^t = \left(\frac{2}{3}, \frac{2}{3}, -\frac{2}{3}\right)^t.
\end{aligned}
$$

Thus, for R^3, $\{(0, 1, 1)^t, (1, -\frac{1}{2}, \frac{1}{2})^t, (\frac{2}{3}, \frac{2}{3}, -\frac{2}{3})^t\}$ is an orthogonal basis, and $\{\frac{1}{\sqrt{2}}(0, 1, 1)^t, \sqrt{\frac{2}{3}}(1, -\frac{1}{2}, \frac{1}{2})^t, \frac{\sqrt{3}}{2}(\frac{2}{3}, \frac{2}{3}, -\frac{2}{3})^t\}$ is an orthonormal basis. Further, from (14.5) it follows that

$$(2, 2, 3)^t = \frac{5}{2}(0, 1, 1)^t + \frac{5/2}{3/2}\left(1, -\frac{1}{2}, \frac{1}{2}\right)^t + \frac{2/3}{4/3}\left(\frac{2}{3}, \frac{2}{3}, -\frac{2}{3}\right)^t.$$

Remark 14.1. From (14.6), we have

$$u^i = (c_{i1}v^1 + \cdots + c_{i,i-1}v^{i-1}) + v^i, \quad i = 1, \cdots, n$$

and hence if S and T are ordered bases, then for changing the basis from T to S the transition matrix is lower triangular and nonsingular as each diagonal element is 1. Then the inverse of this matrix is also lower triangular (see Problem 4.4), i.e., there exist scalars d_{ij}, $i = 1, \cdots, n$, $j = 1, \cdots, i$ such that

$$v^i = d_{i1}u^1 + \cdots + d_{ii}u^i.$$

Remark 14.2. Let (V, C) be an n-dimensional inner product space, and let $u^1, \cdots, u^r$, $r < n$ be orthogonal vectors of V. Then, from Corollary 9.3 and Theorem 14.4 it follows that there are vectors $u^{r+1}, \cdots, u^n$ in V such that $S = \{u^1, \cdots, u^n\}$ is a orthogonal basis of V.

Theorem 14.5. Let (V, C) be an n-dimensional inner product space, and let U be a subspace of V. Then, $V = U \oplus U^\perp$.

Proof. In view of Theorem 14.4, there exists an orthogonal basis $\{u^1, \cdots, u^r\}$ of U, and by Remark 14.2, we can extend it to an orthogonal basis $\{u^1, \cdots, u^n\}$ of V. Clearly, $u^{r+1}, \cdots, u^n \in U^\perp$. Now let $u \in V$, then

$$u = c_1 u^1 + \cdots + c_r u^r + c_{r+1} u^{r+1} + \cdots + c_n u^n,$$

where $c_1 u^1 + \cdots + c_r u^r \in U$ and $c_{r+1} u^{r+1} + \cdots + c_n u^n \in U^\perp$. Thus, $V = U \oplus U^\perp$. On the other hand, if $v \in U \oplus U^\perp$, then $(v, v) = 0$, and this implies that $v = 0$, and hence $U \oplus U^\perp = \{0\}$. Therefore, from Theorem 9.6, we have $V = U \oplus U^\perp$. ∎

Example 14.8. For a given $m \times n$ matrix A if $\{u^1, \cdots, u^r\}$ is a basis (orthogonal basis) of $R(A)$ and $\{u^{r+1}, \cdots, u^n\}$ is a basis (orthogonal basis) for $\mathcal{N}(A)$, then from Theorems 11.4 and 14.5 it immediately follows that $\{u^1, \cdots, u^n\}$ is a basis (orthogonal basis) for R^n. Thus the set of vectors $\{(1, 2, 1, 5)^t, (0, 0, 1, 2)^t, (-2, 1, 0, 0)^t, (-3, 0, -2, 1)^t\}$ obtained in Example 11.1 form a basis of R^4.

Remark 14.3. Theorem 14.5 holds for infinite dimensional inner product spaces also.

Remark 14.4. In view of Theorem 14.5 every $v \in V$ can be uniquely written as $v = u + u'$, where $u \in U$ and $u' \in U^\perp$. We say u is the *orthogonal projection of v along U*, and denote it by $\mathrm{proj}(v, U)$, or $\mathrm{proj}_U v$. In particular, if U is spanned by an orthogonal set $S = \{u^1, \cdots, u^r\}$, then $\mathrm{proj}_S v = \mathrm{proj}_U v = c_1 u^1 + \cdots + c_r u^r$, where $c_i = (v, u^i)/(u^i, u^i)$, $i = 1, \cdots, r$. Thus, $\mathrm{proj}_U v = \sum_{i=1}^r \mathrm{proj}_{u^i} v$. In the case where S is ordered, then $(c_1, \cdots, c_r)^t$ are the coordinates of u with respect to the set S.

Problems

14.1. For the vectors $u = (a_1, a_2, a_3)^t$, $v = (b_1, b_2, b_3)^t \in R^3$ the *cross product* (valid only in R^3), denoted as $u \times v \in R^3$, is defined as

$$u \times v = (a_2 b_3 - a_3 b_2, a_3 b_1 - a_1 b_3, a_1 b_2 - a_2 b_1)^t.$$

Show that for all vectors $u, v, w \in R^3$

(i) $u \times v = -(v \times u)$

(ii) $u \times u = 0$

(iii) the vector $u \times v$ is orthogonal to both u and v

(iv) $u \times (v + w) = (u \times v) + (u \times w)$

(v) $(u \times v) \times w = (u, w)v - (v, w)u$

(vi) $(u \times v)^2 = ((u \times v), (u \times v)) = (u, u)(v, v) - (u, v)^2$

(vii) the absolute value of the *triple product* $(u, (v \times w))$ represents the volume of the parallelepiped formed by the vectors u, v, w.

14.2. Let $u = (a_1, \cdots, a_m)^t \in R^m$, $v = (b_1, \cdots, b_n)^t \in R^n$. The *outer product* of u and v is defined as uv^t, which is an $m \times n$ matrix. For $m = n$ show that $u^t v = \text{tr}(uv^t) = \text{tr}(vu^t)$.

14.3. Let (V, R) be an n-dimensional vector space, and let $S = \{u^1, \cdots, u^n\}$ be an ordered basis for V. If $u, v \in V$ are such that $u = a_1 u^1 + \cdots + a_n u^n$ and $v = b_1 u^1 + \cdots + b_n u^n$, then show that $(u, v) = (y_S(u), y_S(v)) = \sum_{i=1}^{n} a_i b_i$ is an inner product on V.

14.4. Let (V, R) be an n-dimensional inner product space, and let $S = \{u^1, \cdots, u^n\}$ be an ordered basis for V. Show that

(i) the *matrix of the inner product* $C = (c_{ij})$, where $c_{ij} = (u^i, u^j)$ is symmetric

(ii) if $a = (a_1, \cdots, a_n)^t$ and $b = (b_1, \cdots, b_n)^t$ are the coordinates of the vectors $u, v \in V$, then $(u, v) = a^t C b$, i.e., the matrix C determines (u, v) for every u and v in V.

14.5. Let V and W be finite dimensional inner product spaces with inner products $(\cdot, \cdot)_V$ and $(\cdot, \cdot)_W$, and let $T : V \to W$ be a linear mapping. The mapping (if it exists) $T^* : W \to V$ is called the *adjoint mapping* of T if for all $u \in V$ and $w \in W$, $(T(u), w)_W = (u, T^*(w))_V$. Show that

(i)   if $T = A \in R^{m \times n}$, then $T^* = A^t$

(ii) if $T = A \in C^{m \times n}$, then $T^* = A^H$

(iii) $(T^*)^* = T$.

14.6. Find a nonzero vector v that is orthogonal to the given vectors

(i) $(1, 3, 5)^t$, $(3, 5, 1)^t$, $v \in R^3$

(ii) $(1,1,2,2)^t$, $(1,3,3,1)^t$, $(5,5,1,1)^t$, $v \in R^4$.

14.7. Find a basis for the subspace $u^\perp$ of R^3, where

(i) $u = (1,3,5)^t$
(ii) $u = (-1,0,1)^t$.

14.8. Let U, W be subsets of an inner product space V. Show that

(i) $U \subseteq U^{\perp\perp}$
(ii)   if $U \subseteq W$, then $W^\perp \subseteq U^\perp$
(iii) $U^\perp = \text{span}(U)^\perp$.

14.9. Let U, W be subspaces of a finite dimensional inner product space V. Show that

(i) $U = U^{\perp\perp}$
(ii) $(U + W)^\perp = U^\perp \cap W^\perp$
(iii) $(U \cap W)^\perp = U^\perp + W^\perp$.

14.10. Let $\{\tilde{u}^1, \cdots, \tilde{u}^n\}$ and $\{\tilde{v}^1, \cdots, \tilde{v}^n\}$ be orthonormal ordered basis of an n-dimensional real inner product space V. Show that the transition matrix A defined in (10.4) is orthogonal.

14.11. Let A be a real (complex) square matrix. Show that the following are equivalent

(i) A is orthogonal (unitary)
(ii)   the rows of A form an orthonormal set
(iii) the columns of A form an orthonormal set.

14.12. For the matrices given in Problem 11.1 use the method of Example 14.8 to find bases of R^5 and R^6.

14.13. For the given basis for R^3 use the Gram–Schmidt orthogonalization process to find an orthonormal basis for R^3

(i) $\{(1,1,1)^t, (-1,1,0)^t, (-1,0,1)^t\}$
(ii) $\{(1,0,1)^t, (0,-1,1)^t, (0,-1,-1)^t\}$.

14.14. Enlarge the following sets of linearly independent vectors to orthonormal bases of R^3 and R^4

(i) $\{(1,1,1)^t, (1,1,2)^t\}$
(ii) $\{(1,1,1,3)^t, (1,2,3,4)^t, (2,3,4,9)^t\}$.

14.15. Show that the set

$$\left\{\sqrt{\frac{2}{\pi}}\sin nx, \quad n=1,2,\cdots\right\}$$

is orthonormal on $0 < x < \pi$. This set generates the Fourier sine series.

14.16. Show that the set

$$\left\{\frac{1}{\sqrt{\pi}}, \sqrt{\frac{2}{\pi}}\cos nx, \quad n=1,2,\cdots\right\}$$

is orthonormal on $0 < x < \pi$. This set generates the Fourier cosine series.

14.17. Show that the set

$$\left\{\frac{1}{\sqrt{2\pi}}, \frac{1}{\sqrt{\pi}}\cos nx, \frac{1}{\sqrt{\pi}}\sin nx, \quad n=1,2,\cdots\right\}$$

is orthonormal on $-\pi < x < \pi$. This set generates the Fourier trigonometric series.

14.18. *Legendre polynomials*, denoted as $P_n(x)$, $n=0,1,2,\cdots$ can be defined by *Rodrigues' formula*

$$P_n(x) = \frac{1}{2^n\,n!}\frac{d^n}{dx^n}(x^2-1)^n, \quad n=0,1,2,\cdots. \tag{14.7}$$

In fact, from (14.7) we easily obtain

$$P_0(x) = 1, \quad P_1(x) = x, \quad P_2(x) = \frac{1}{2}(3x^2-1), \quad P_3(x) = \frac{1}{2}(5x^3-3x),$$

$$P_4(x) = \frac{1}{8}(35x^4-30x^2+3), \quad P_5(x) = \frac{1}{8}(63x^5-70x^3+15x),\cdots.$$

Show that the set $\{P_0(x), P_1(x), P_2(x),\cdots\}$ is orthogonal in the interval $[-1,1]$.

14.19. Consider the space V of all real infinite sequences $u = (a_1, a_2,\cdots)$ satisfying $\sum_{i=1}^{\infty} a_i^2 < \infty$. Addition and scalar multiplication for all $u = (a_1, a_2,\cdots), v = (b_1, b_2,\cdots) \in V$, $c \in R$ is defined as $(u+v) = (a_1+b_1, a_2+b_2,\cdots)$, $cu = (ca_1, ca_2,\cdots)$. Show that

(i) V is a vector space

(ii)   the inner product $(u, v) = a_1b_1 + a_2b_2 + \cdots$ is well defined, i.e., $\sum_{i=1}^{\infty} a_ib_i$ converges absolutely.

This inner product space is called ℓ_2-*space* and is an example of a *Hilbert space*.

14.20. Let $u = (a_1, \cdots, a_n)^t, v = (b_1, \cdots, b_n)^t \in R^n$, and let $w_1, \cdots, w_n$ be fixed real numbers. Show that

$$(u, v) = w_1 a_1 b_1 + w_2 a_2 b_2 + \cdots + w_n a_n b_n$$

defines an inner product (known as *weighted inner product*) in R^n.

Answers or Hints

14.1. Verify directly.

14.2. Verify directly.

14.3. Verify directly.

14.4. (i) $(u^i, u^j) = (u^j, u^i)$

(ii) $(u, v) = \sum_{i=1}^n \sum_{j=1}^n a_i c_{ij} b_j$.

14.5. See (14.3).

14.6. (i) $(11, -7, 2)^t$.

(ii) $(-1, 1, -1, 1)^t$.

14.7. (i) $\{(-3, 1, 0)^t, (5, 0, -1)^t\}$.

(ii) $\{(1, 0, 1)^t, (0, 1, 0)^t\}$.

14.8. (i) Let $u \in U$. Then $(u, v) = 0$ for every $v \in U^\perp$. Hence $u \in U^{\perp\perp}$, and therefore, $U \subseteq U^{\perp\perp}$.

(ii) Let $u \in W^\perp$. Then $(u, v) = 0$ for every $v \in W$. Since $U \subseteq W$, $(u, v) = 0$ for every $v \in U$. Thus $u \in U^\perp$, and hence $W^\perp \subseteq U^\perp$.

(iii) Clearly, $U \subseteq \text{span}(U)$, and hence from (ii), $\text{span}(U)^\perp \subseteq U^\perp$. If $u \in U^\perp$ and $v \in \text{span}(U)$, then there exist $v^1, \cdots, v^r \in U$ such that $v = c_1 v^1 + \cdots + c_r v^r$, but then $(u, v) = (u, c_1 v^1 + \cdots + c_r v^r) = c_1(u, v^1) + \cdots + c_r(u, v^r) = c_1(0) + \cdots + c_r(0) = 0$. Thus $u \in \text{span}(U)^\perp$, i.e., $U^\perp \subseteq \text{span}(U)^\perp$.

14.9. (i) From Theorem 14.5, $V = U \oplus U^\perp$ and $V = U^\perp \oplus U^{\perp\perp}$. Thus from Corollary 9.4, we have $\text{Dim } U = \text{Dim } U^{\perp\perp}$. Further, since from Problem 14.8(i), $U \subseteq U^{\perp\perp}$, it follows that $U = U^{\perp\perp}$.

14.10. From (10.1) we have $\delta_{ij} = (\tilde{v}^i, \tilde{v}^j) = (\sum_{k=1}^n a_{ki} \tilde{u}^k, \sum_{\ell=1}^n a_{\ell j} \tilde{u}^\ell) = \sum_{k=1}^n a_{ki} a_{kj} = c_{ij}$. Now note that $A^t A = (c_{ij}) = I$.

14.11. Recall the definition of an orthogonal (unitary) matrix in Problem 4.7.

14.12. (i) $\{(1, 0, -1, -2, 0)^t, (0, 1, 2, 3, 0)^t, (0, 0, 0, 0, 1)^t, (2, -3, 0, 1, 0)^t, (1, -2, 1, 0, 0)^t\}$.

(ii) $\left\{ \left(1, 3, 0, 0, \frac{5}{4}, \frac{5}{2}\right)^t, \left(0, 0, 1, 0, -\frac{83}{16}, -\frac{41}{8}\right)^t, \left(0, 0, 0, 1, \frac{61}{16}, \frac{31}{8}\right)^t, (-3, 1, 0, 0, 0, 0)^t, \left(-\frac{5}{4}, 0, \frac{83}{16}, -\frac{61}{16}, 1, 0\right)^t, \left(-\frac{5}{2}, 0, \frac{41}{8}, -\frac{31}{8}, 0, 1\right)^t \right\}$.

14.13. (i) $\{\frac{1}{\sqrt{3}}(1, 1, 1)^t, \frac{1}{\sqrt{2}}(-1, 1, 0)^t, \frac{1}{\sqrt{6}}(-1, -1, 2)^t\}$.

(ii) $\{\frac{1}{\sqrt{2}}(1, 0, 1)^t, \frac{1}{\sqrt{6}}(1, 2, -1)^t, \frac{1}{\sqrt{3}}(1, -1, -1)^t\}$.

14.14. (i) $\left\{ \frac{1}{\sqrt{3}}(1, 1, 1), \frac{1}{\sqrt{6}}(-1, -1, 2), \frac{1}{\sqrt{2}}(1, 1, 0) \right\}$.

(ii) $\left\{ \frac{1}{2\sqrt{3}}(1, 1, 1, 3)^t, \frac{1}{2\sqrt{3}}(-1, 1, 3, -1)^t, \frac{1}{\sqrt{6}}(-2, -1, 0, 1)^t, \frac{1}{\sqrt{6}}(1, -2, 1, 0)^t \right\}$.

14.15. See Problem 8.10(i).

14.16. See Problem 8.10(ii).

14.17. See Problem 8.10(iii).

14.18. From (14.7), we have $2^n n! \int_{-1}^{1} P_m(x) P_n(x) dx = \int_{-1}^{1} P_m(x) \frac{d^n}{dx^n}(x^2 - 1)^n dx$. Now an integration by parts gives $\int_{-1}^{1} P_m(x) \frac{d^n}{dx^n}(x^2 - 1)^n dx = P_m(x) \times \frac{d^{n-1}}{dx^{n-1}}(x^2 - 1)^n |_{-1}^{1} - \int_{-1}^{1} \frac{d}{dx} P_m(x) \frac{d^{n-1}}{dx^{n-1}}(x^2 - 1)^n dx$. However, since $d^{n-1}(x^2 -1)^n/dx^{n-1}$ contains a factor $(x^2 - 1)$, it follows that $2^n n! \int_{-1}^{1} P_m(x) P_n(x) dx = -\int_{-1}^{1} \frac{d}{dx} P_m(x) \frac{d^{n-1}}{dx^{n-1}}(x^2 -1)^n dx$. We can integrate the right side once again, and continue until we have performed n such integrations. At this stage, we find $2^n n! \int_{-1}^{1} P_m(x) P_n(x) dx = (-1)^n \int_{-1}^{1} \left(\frac{d^n}{dx^n} P_m(x)\right)(x^2 -1)^n dx$. There is no loss of generality if we assume that $m \leq n$. If $m < n$, then $d^n P_m(x)/dx^n = 0$ and it follows that $\int_{-1}^{1} P_m(x) P_n(x) dx = 0$.

14.19. In $\sum_{i=1}^{\infty}(\alpha a_i + \beta b_i)^2 = \alpha^2 \sum_{i=1}^{\infty} a_i^2 + \alpha\beta \sum_{i=1}^{\infty} a_i b_i + \beta^2 \sum_{i=1}^{\infty} b_i^2$ use $a_i b_i \leq (1/2)(a_i^2 + b_i^2)$.

14.20. Verify directly.

Chapter 15

Linear Functionals

Let V be a vector space over the field F. A linear mapping $\phi : V \to F$ is called a *linear functional on V*. Since a linear functional is a special type of linear mapping, all the results presented in Chapters 11 and 12 for general mappings hold for linear functionals also. Therefore, in this chapter we shall present only those results that have special significance for linear functionals. We begin with the following interesting examples.

Example 15.1. Let $V = F^n$. Then, for $u = (u_1, \cdots, u_n) \in V$ the projection mapping $\phi_i(u_1, \cdots, u_n) = u_i$ is a linear functional.

Example 15.2. Let $V = C[a, b]$ be the space of all continuous real–valued functions on the interval $[a, b]$. Then, for $f \in V$ the integral mapping $\phi(f) = \int_a^b f(x)dx$ is a linear functional. The mapping $\phi(f) = f(x_0)$, where $x_0 \in [a, b]$, but fixed, is also a linear functional.

Example 15.3. Let $V = M^{n \times n}$. Then, for $A = (a_{ij}) \in V$ the trace mapping $\phi(A) = \text{tr}(A)$ is a linear functional.

Example 15.4. In the inner product space (V, C), let the vector u^0 be fixed. Then, for $u \in V$ the mapping (u, u_0) is a linear functional; however, (u_0, u) is not a linear functional because $(u_0, \alpha u) = \overline{\alpha}(u_0, u)$.

The vector space $\text{Hom}(V, F)$ is called the *dual space* of V, and is denoted as V^*. In view of Theorem 12.5 it is clear that $\dim V = \dim V^*$.

Theorem 15.1. Let $S = \{u^1, \cdots, u^n\}$ be a basis of V, and let $\phi_1, \cdots, \phi_n \in V^*$ be linear functionals defined by $\phi_j(u^i) = \delta_{ij}$. Then, $S^* = \{\phi_1, \cdots, \phi_n\}$ is a basis (called *dual basis*) of V^*.

Proof. First we shall show that S^* spans V^*. For this, let $\phi \in V^*$ and suppose that $\phi(u^i) = c_i$, $i = 1, \cdots, n$. We set $\psi = c_1\phi_1 + \cdots + c_n\phi_n$. Then, we have $\psi(u^i) = c_1\phi(u^i) + \cdots + c_n\phi_n(u^i) = c_i = \phi(u^i)$, $i = 1, \cdots, n$. Thus, ψ and ϕ have the same values on the basis S, and hence must be the same on V. Therefore, S^* spans V^*. To show that S^* is linearly independent, let $a_1\phi_1 + \cdots + a_n\phi_n = 0$. Then, we have $0 = 0(u_i) = a_1\phi_1(u^i) + \cdots + a_n\phi(u^i) = a_i$, $i = 1, \cdots, n$ as required. ∎

Remark 15.1. Let $0 \neq v \in V$, and extend $\{v\}$ to a basis $\{v, v_2, \cdots, v_n\}$ of

V. Then, from Theorem 15.1 there exists a unique linear mapping $\phi : V \to F$ such that $\phi(v) = 1$ and $\phi(v_i) = 0$, $i = 2, \cdots, n$.

Example 15.5. From Example 6.1, we know that $S = \{u^1 = (2, -1, 1),$ $u^2 = (3, 2, -5),$ $u^3 = (1, 3, -2)\}$ is a basis of R^3. We shall find its dual basis $S^* = \{\phi_1, \phi_2, \phi_3\}$. We let $\phi_1(x, y, z) = a_{11}x + a_{12}y + a_{13}z$, $\phi_2(x, y, z) = a_{21}x + a_{22}y + a_{23}z$, $\phi_3(x, y, z) = a_{31}x + a_{32}y + a_{33}z$. Since $\phi_j(u^i) = \delta_{ij}$, we need to solve the systems

$$
\begin{array}{rclcrcl}
2a_{11} - a_{12} + a_{13} & = & 1 & \quad & 2a_{21} - a_{22} + a_{23} & = & 0 \\
3a_{11} + 2a_{12} - 5a_{13} & = & 0 & & 3a_{21} + 2a_{22} - 5a_{23} & = & 1 \\
a_{11} + 3a_{12} - 2a_{13} & = & 0, & & a_{21} + 3a_{22} - 2a_{23} & = & 0
\end{array}
$$

and

$$
\begin{array}{rcl}
2a_{31} - a_{32} + a_{33} & = & 0 \\
3a_{31} + 2a_{32} - 5a_{33} & = & 0 \\
a_{31} + 3a_{32} - 2a_{33} & = & 1.
\end{array}
$$

Now in view of Example 6.6, solutions of these systems can be written as $(a_{11}, a_{12}, a_{13}) = (11/28, 1/28, 7/28)$, $(a_{21}, a_{22}, a_{23}) = (1/28, -5/28, -7/28)$, and $(a_{31}, a_{32}, a_{33}) = (3/28, 13/28, 7/28)$. Thus, it follows that

$$
\begin{aligned}
\phi_1(x, y, z) & = \frac{11}{28}x + \frac{1}{28}y + \frac{7}{28}z \\
\phi_2(x, y, z) & = \frac{1}{28}x - \frac{5}{28}y - \frac{7}{28}z \\
\phi_3(x, y, z) & = \frac{3}{28}x + \frac{13}{28}y + \frac{7}{28}z.
\end{aligned}
$$

Theorem 15.2. Let $S = \{u^1, \cdots, u^n\}$ and $T = \{v^1, \cdots, v^n\}$ be bases of V, and let $S^* = \{\phi_1, \cdots, \phi_n\}$ and $T^* = \{\psi_1, \cdots, \psi_n\}$ be the corresponding dual bases of V^*. Further, let $A = (a_{ij})$ be the transition (change-of-basis) matrix from T to S, and $B = (b_{ij})$ be the change-of-basis matrix from T^* to S^*. Then, $A^t B = I$, i.e., $B = (A^{-1})^t$.

Proof. In view of (10.1), we have

$$
v^i = a_{1i}u^1 + a_{2i}u^2 + \cdots + a_{ni}u^n
$$

and

$$
\psi_j = b_{1j}\phi_1 + b_{2j}\phi_2 + \cdots + b_{nj}\phi_n.
$$

Thus, it follows that

$$
\begin{aligned}
\delta_{ij} = \psi_j(v^i) & = \sum_{k=1}^{n} b_{kj}\phi_k \left(\sum_{\ell=1}^{n} a_{\ell i}u^i \right) = \sum_{k=1}^{n} b_{kj}a_{ki} \\
& = (a_{1i}, \cdots, a_{ni})(b_{1j}, \cdots, b_{nj})^t.
\end{aligned}
$$

This immediately gives $A^t B = I$. ∎

Since V^* is a vector space, it has a dual space, denoted as V^{**}, and called the *second dual* of V. Thus, V^{**} is a collection of all linear functionals on V^*. It follows that corresponding to each $v \in V$ there is a distinct $\hat{v} \in V^{**}$. To show this, for any $\phi \in V^*$, we define $\hat{v}(\phi) = \phi(v)$. Now it suffices to show that the map $\hat{v} : V^* \to F$ is linear. For this, we note that for scalars $a, b \in F$ and linear functionals $\phi, \psi \in V^*$, we have

$$\hat{v}(a\phi + b\psi) = (a\phi + b\psi)(v) = a\phi(v) + b\psi(v) = a\hat{v}(\phi) + b\hat{v}(\psi).$$

Theorem 15.3. Let V be a finite dimensional vector space over the field F, then the mapping $v \mapsto \hat{v}$, known as *natural mapping*, is an isomorphism of V onto V^{**}.

Proof. For any $v, w \in V$ and $a, b \in F$, and $\phi \in V^*$, we have

$$(\widehat{av + bw})(\phi) = \phi(av + bw) = a\phi(v) + bw(b) = a\hat{v}(\phi) + b\hat{w}(\phi) = (a\hat{v} + b\hat{w})(\phi),$$

and hence the mapping $v \mapsto \hat{v}$ is linear. Further, from Remark 15.1, for every $0 \neq v \in V$ there exists $\phi \in V^*$ so that $\phi(v) \neq 0$. This implies that $\hat{v}(\phi) = \phi(v) \neq 0$, and hence $\hat{v} \neq 0$. Thus, we can conclude that the mapping $v \mapsto \hat{v}$ is nonsingular, which in turn shows that it is an isomorphism. ∎

Now let W be a subset (not necessarily subspace) of the vector space V over the field F. The *annihilator* of W is the set W^0 of linear functionals $\phi \in V^*$ such that $\phi(w) = 0$ for every $w \in W$. It follows rather easily that $W^0 \subseteq V^*$ is a subspace; if $\phi \in V^*$ annihilates W, then ϕ annihilates $\text{Span}(W)$, i.e., $W^0 = [\text{Span}(W)]^0$; if $W = \{0\}$ then $W^0 = V^*$, and if $W = V$ then W^0 is the null space of V^*. We also define $W^{00} = \{u \in V : \varphi(u) = 0 \text{ for every } \varphi \in W^0\}$.

Theorem 15.4. Let V be a finite dimensional vector space over the field F, and let W be a subspace of V. Then, $\dim W + \dim W^0 = \dim V$, and $W^{00} = W$.

Proof. Suppose that $\dim V = n$ and $\dim W = r$. We need to show that $\dim W^0 = n - r$. Let $\{w^1, \cdots, w^r\}$ be a basis of W. We extend it so that $\{w^1, \cdots, w^r, u^1, \cdots, u^{n-r}\}$ is a basis of V. Let $\{\phi_1, \cdots, \phi_r, \varphi_1, \cdots, \varphi_{n-r}\}$ be the basis of V^*, which is dual to this basis of V. Now by the definition of the dual basis, each φ_j annihilates each w^i, and hence $\varphi_1, \cdots, \varphi_{n-r} \in W^0$. It suffices to show that $\Phi = \{\varphi_1, \cdots, \varphi_{n-r}\}$ is a basis of W^0. For this, since Φ is a subset of a basis of V^*, it is linearly independent. To show Φ spans W^0, let $\varphi \in W^0$, then in view of Problem 15.4, we have

$$
\begin{aligned}
\varphi &= \varphi(w^1)\phi_1 + \cdots + \varphi(w^r)\phi_r + \varphi(u^1)\varphi_1 + \cdots \varphi(u^{n-r})\varphi_{n-r} \\
&= \varphi(u^1)\varphi_1 + \cdots \varphi(u^{n-r})\varphi_{n-r}.
\end{aligned}
$$

Example 15.6. We shall find a basis of the annihilator W^0 of the subspace

W of R^4 spanned by $w^1 = (1, 2, 0, -1)$ and $w^2 = (5, 2, 4, 9)$. Since $W^0 = [\text{Span}(W)]^0$, it suffices to find a basis of the set of linear functionals ϕ such that $\phi(w^1) = 0$ and $\phi(w^2) = 0$, where $\phi(w_1, w_2, w_3, w_4) = aw_1 + bw_2 + cw_3 + dw_4$. For this, the system

$$\begin{aligned} \phi(w^1) &= \phi(1, 2, 0, -1) &= a + 2b + 0c - d &= 0 \\ \phi(w^2) &= \phi(5, 2, 4, 9) &= 5a + 2b + 4c + 9d &= 0 \end{aligned}$$

has solutions with c and d as free variables. We fix $c = 1$ and $d = 0$, to get $a = -1$ and $b = 1/2$. Next, we fix $c = 0$ and $d = 1$, to get $a = -5/2$ and $b = 7/4$. Hence, the following linear functions

$$\phi_1(w^1) = -w_1 + \frac{1}{2}w_2 + w_3 \quad \text{and} \quad \phi_2(w^2) = -\frac{5}{2}w_1 + \frac{7}{4}w_2 + w_4$$

form the basis of W^0. Similarly, a basis of the annihilator W^0 of the subspace W of R^4 spanned by $w^1 = (1, 2, 0, -1)$, $w^2 = (5, 2, 4, 9)$, and $w^3 = (-2, 2, -3, 2)$ is the linear function $\phi(w^i) = -2w_1 + w_2 + 2w_3$, $i = 1, 2, 3$.

Now we shall prove the following important result.

Theorem 15.5 (Riesz representation theorem). Let V be a finite dimensional vector space over the field F $(F = R, C)$ on which $(\cdot, \cdot)$ is an inner product. Let $f : V \to F$ be a linear functional on V. Then, there exists a vector $u \in V$ such that $f(v) = (v, u)$ for all $v \in V$.

Proof. Using the Gram–Schmidt orthogonalization process we can find an orthonormal basis of V, say, $\{v^1, \cdots, v^n\}$. Now for an arbitrary vector $v \in V$, we have $v = (v, v^1)v^1 + \cdots + (v, v^n)v^n$. Thus, it follows that $f(v) = (v, v^1)f(v^1) + \cdots + (v, v^n)f(v^n) = (v, v^1\overline{f(v^1)} + \cdots + v^n\overline{f(v^n)})$. Denoting by $u = v^1\overline{f(v^1)} + \cdots + v^n\overline{f(v^n)}$, the result follows. ∎

Remark 15.2. The vector u in Theorem 15.5 is unique. Indeed, if there are two vectors u^1, u^2 such that $f(v) = (v, u^1) = (v, u^2)$, then $(v, u^1 - u^2) = 0$ for all $v \in V$. But then for $v = u^1 - u^2 \in V$ it follows that $0 = (u^1 - u^2, u^1 - u^2) = \|u^1 - u^2\|^2 = 0$, which implies that $u^1 = u^2$.

Let V and W be two vector spaces on which inner products are defined. Let $f : V \to W$ be a linear mapping. For each vector $w \in W$, we define a map $f_w : V \to R$ as $f_w(v) = (f(v), w)$. It follows that this map is a linear functional. Indeed, we have $f_w(v^1 + v^2) = (f(v^1 + v^2), w) = (f(v^1) + f(v^2), w) = (f(v^1), w) + (f(v^2), w) = f_w(v^1) + f_w(v^2)$, also $f_w(\alpha v) = (f(\alpha v), w) = (\alpha f(v), w) = \alpha(f(v), w) = \alpha f_w(v)$.

Example 15.7. Let $f : R^2 \to R^3$ given by $f(v_1, v_2) = (v_1 - 2v_2, -7v_1, 4v_1 - 9v_2)$. Let $w = (-3, 1, 2) \in R^3$. Then, we have $f_w(v) = -3(v_1 - 2v_2) + 1(-7v_1) + 2(4v_1 - 9v_2) = -2v_1 - 12v_2$.

In view of above notations and Riesz's Theorem for the linear functional

f_w, there is a vector in V denoted as $f^*(w)$ such that $f_w(v) = (v, f^*(w))$. Thus, it follows that $(f(v), w) = (v, f^*(w))$ for all $v \in V$ and $w \in W$.

From Remark 15.2, the map $f^* : W \to V$ is well defined and is called the *adjoint* of the linear mapping f.

Theorem 15.6. If $f : V \to W$ is a linear mapping, then the adjoint map $f^* : W \to V$ is also a linear mapping.

Proof. For all $v \in V$, we have $(v, f^*(w^1 + w^2)) = (f(v), w^1 + w^2) = (f(v), w^1) + (f(v), w^2) = (v, f^*(w^1)) + (v, f^*(w^2)) = (v, f^*(w^1) + f^*(w^2))$. But then from Remark 15.2 it follows that $f^*(w^1 + w^2) = f^*(w^1) + f^*(w^2)$. Now for $\alpha \in R$ and all $v \in V$, we have $(v, f^*(\alpha w)) = (f(v), \alpha w) = \alpha(f(v), w) = \alpha(v, f^*(w)) = (v, \alpha f^*(w))$, which implies that $f^*(\alpha w) = \alpha f^*(w)$. ∎

Example 15.8. We consider the same linear mapping as in Example 15.7, and let $(a, b, c) \in R^3$. Then, from $(f(v_1, v_2), (a, b, c)) = ((v_1 - 2v_2, -7v_1, 4v_1 - 9v_2), (a, b, c)) = a(v_1 - 2v_2) + b(-7v_1) + c(4v_1 - 9v_2) = v_1(a - 7b + 4c) + v_2(-2a - 9c) = ((v_1, v_2), (a - 7b + 4c, -2a - 9c))$, we have $f^*(a, b, c) = (a - 7b + 4c, -2a - 9c)$.

Problems

15.1. For the given basis $S = \{(11, 19, 21), (3, 6, 7), (4, 5, 8)\}$ of R^3, find its dual basis.

15.2. For the given basis $S = \{(1, 2, 1, 5), (0, 0, 1, 2), (-2, 1, 0, 0), (-3, 0, -2, 1)\}$ of R^4, find its dual basis.

15.3. For $f \in (P_3, R)$ let the linear functionals $\phi_{-1}, \phi_0, \phi_1$ be defined by $\phi_{-1}(f(x)) = f(-1)$, $\phi_0(f(x)) = f(0)$, $\phi_1(f(x)) = f(1)$. Show that $S^* = \{\phi_{-1}, \phi_0, \phi_1\}$ is linearly independent, and find the basis S of (P_3, R) so that S^* is its dual.

15.4. Repeat Problem 15.3 with $\phi_{-1}(f(x)) = \int_{-1}^0 f(x)dx$, $\phi_0(f(x)) = f(0)$, $\phi_1(f(x)) = \int_0^1 f(x)dx$.

15.5. Let V, S and S^* be as in Theorem 15.1. Then, show that for any $u \in V$, $u = \sum_{i=1}^n \phi_i(u)u^i$, and for any $\phi \in V^*$, $\phi = \sum_{i=1}^n \phi(u^i)\phi_i$.

15.6. Let (V, R) be a real inner product space. For each $u \in V$ we define the linear functional as $L_u(v) = (u, v)$ for all $v \in V$. Show that

(i) the set of all linear functionals L_u, denoted as $\tilde{V}$, is a linear space

(ii) $L_u = 0$ if and only if $u = 0$

(iii) for a given functional $\phi \in V^*$ there exists a unique vector $u \in V$ such that $\phi(v) = (u, v)$ for all $v \in V$

(iv) the mapping $u \to L_u$ of V into V^* is an isomorphism.

15.7. Let $f : R^3 \to R$, $f(v) = 2v_1 - 4v_2 + 5v_3$ and $g : R^5 \to R$, $g(v) = 5v_1 - 7v_2 + 8v_3 - 5v_4 + v_5$ be two linear functionals. Find the vector $u \in R^3$ such that $f(v) = (v, u)$, and the vector $w \in R^5$ such that $g(v) = (v, w)$.

15.8. Let $f : R^3 \to R^3$, $f(v) = (v_1 - 2v_2 + v_3, -2v_1 + v_3, -v_2 + v_3)$. For $w = (4, -1, 7)$ find $f_w(v) = (f(v), w)$.

15.9. Let $f : R^4 \to R^3$, $f(v) = (5v_1 - 2v_2 + 4v_3, 4v_1 + 6v_3 + v_4, -v_2 + v_4)$.

(i)   Find the adjoint map $f^* : R^3 \to R^4$.

(ii)   Find the matrices A and B of the maps f and f^*, and note that $A = B^t$.

(iii)   Generalize part (ii) for arbitrary vector spaces V and W.

15.10. Show that Theorem 15.5 does not hold for spaces of infinite dimensions.

Answers or Hints

15.1. $\phi_1 = \frac{1}{30}(13x+4y-9z)$, $\phi_2 = \frac{1}{30}(-47x+4y+21z)$, $\phi_3 = \frac{1}{30}(7x-14y+9z)$.
15.2. $\phi_1 = \frac{1}{34}(5x + 10y - 6z + 3w)$, $\phi_2 = \frac{1}{34}(-11x - 22y + 20z + 7w)$, $\phi_3 = \frac{1}{34}(-10x + 14y + 12z - 6w)$, $\phi_4 = \frac{1}{34}(-3x - 6y - 10z + 5w)$.
15.3. $S = \left\{\frac{1}{2}x(x - 1), -(x + 1)(x - 1), \frac{1}{2}x(x + 1)\right\}$.
15.4. $S = \left\{\frac{3}{2}x\left(x - \frac{2}{3}\right), -3\left(x^2 - \frac{1}{3}\right), \frac{3}{2}x\left(x + \frac{2}{3}\right)\right\}$.
15.5. If $u = \sum_{i=1}^n c_i u^i$, then $\phi_i(u) = c_i$. Now for all u, $\phi(u) = \sum_{i=1}^n \phi_i(u) \times \phi(u^i) = \left(\sum_{i=1}^n \phi(u^i)\phi_i\right)(u)$.
15.6. (i) From the definition of real inner product $L_{c_1u+c_2w} = c_1 L_u + c_2 L_w$ for all $u, w \in V$ and $c_1, c_2 \in R$.
(ii)   Follows from the definition of real inner product.
(iii)   If $S = \{u^1, \cdots, u^n\}$ is a basis of V, then $\tilde{S} = \{L_{u^1}, \cdots, L_{u^n}\}$ is a basis of $\tilde{V}$. Indeed, if $L_u \in \tilde{V}$ and $u = c_1 u^1 + \cdots + c_n u^n$, then $L_u = L_{c_1 u^1 + \cdots + c_n u^n} = c_1 L_{u^1} + \cdots + c_n L_{u^n}$. Also, if $a_1 L_{u^1} + \cdots + a_n L_{u^n} = 0$, then $L_{a_1 u^1 + \cdots + a_n u^n} = 0$, which in view of Part (ii) implies that $a_1 u^1 + \cdots + a_n u^n = 0$. But, then $a_1 = \cdots = a_n = 0$. Thus, $\tilde{V}$ is a subspace of V^*, and since $\dim V^* = \dim \tilde{V}$, $V^* = \tilde{V}$. Now if $\phi \in V^*$, then there exist unique real numbers b_i, $i = 1, \cdots, n$ such that $\phi = b_1 L_{u^1} + \cdots + b_n L_{u^n} = L_{b_1 u^1 + \cdots + b_n u^n}$.
(iv)   Use Theorem 12.2.
15.7. $u = (2, -4, 5)$, $w = (5, -7, 8, -5, 1)$.
15.8. $f_w(v) = 6v_1 - 15v_2 + 10v_3$.
15.9. (i) $f^*(a, b, c) = (5a + 4b, -2a - c, 4a + 6b, b + c)$.

(ii) $A = \begin{pmatrix} 5 & -2 & 4 & 0 \\ 4 & 0 & 6 & 1 \\ 0 & -1 & 0 & 1 \end{pmatrix}$, $B = \begin{pmatrix} 5 & 4 & 0 \\ -2 & 0 & -1 \\ 4 & 6 & 0 \\ 0 & 1 & 1 \end{pmatrix}$.

(iii)   Let V and W be complex vector spaces, $f : V \to W$ be a linear map with associated matrix A, and $f^* : W \to V$ its adjoint with associated matrix B. We need to show that $A = \overline{B}^t$. For this, we assume that $\dim V = n$ and $\dim W = m$, and $\{v^1, \cdots, v^n\}$, $\{w^1, \cdots, w^m\}$ are their orthonormal bases. Clearly, $A = (a_{ij}) = ((f(v^i), w^j))$, $1 \le i \le n$, $1 \le j \le m$, and $B = (b_{ij}) = ((f^*(w^i), v^j))$, $1 \le i \le m$, $i \le j \le n$. Since $(f^*(w^i), v^j) = \overline{(v^j, f^*(w^i))} = \overline{(f(v^j), w^i)}$, it follows that $a_{ij} = \overline{b}_{ji}$.

15.10. Let V be the space of all polynomials over R with the inner product $(f(x), g(x)) = \int_0^1 f(x)g(x)dx$, i.e., if $f(x) = \sum_{i=0}^n a_i x^i$ and $g(x) = \sum_{j=0}^m b_j x^j$, then $(f(x), g(x)) = \sum_{i=0}^n \sum_{j=0}^m (i + j + 1)^{-1} a_i b_j$. We consider $\phi : V \to R$ as $\phi(f) = f(0)$. We shall show that a polynomial $g(x)$ such that $f(0) = \phi(f) = (f(x), g(x)) = \int_0^1 f(x)g(x)dx$ does not exist for every $f(x)$. For this, note that for any $f(x)$ we have $\phi(xf(x)) = 0$. Thus, for any $f(x)$, $0 = \int_0^1 xf(x)g(x)dx$. Hence, in particular for $f(x) = xg(x)$, we have $0 = \int_0^1 x^2 g^2(x)dx$, which implies that $g(x) = 0$, and therefore $\phi(f(x)) = (f(x), 0) = 0$. But this contradicts the fact that ϕ is a nonzero functional.

Chapter 16

Eigenvalues and Eigenvectors

Eigenvalues and eigenvectors of a matrix are of great importance in the qualitative as well as quantitative study of many physical problems. For example, stability of an aircraft is determined by the location of the eigenvalues of a certain matrix in the complex plane. Basic solutions of ordinary differential and difference equations with constant coefficients are expressed in terms of eigenvalues and eigenvectors of the coefficient matrices. In this chapter we shall mainly summarize properties of the eigenvalues and eigenvectors of matrices, which are of immense value.

The number λ, real or complex, is called an *eigenvalue (characteristic root, latent root, proper value)* of an $n \times n$ matrix A if there exists a nonzero real or complex vector u such that $Au = \lambda u$, i.e., Au is parallel to u. The vector u is called an *eigenvector*, corresponding to the eigenvalue λ. From Corollary 5.1, λ is an eigenvalue of A if and only if it is a solution of the *characteristic equation*

$$p(\lambda) = \det(A - \lambda I) = 0, \qquad (16.1)$$

which in the expanded form, known as *characteristic polynomial* of A, can be written as

$$p(\lambda) = (-1)^n \lambda^n + a_1 \lambda^{n-1} + \cdots + a_{n-1}\lambda + a_n = 0. \qquad (16.2)$$

Thus from the *fundamental theorem of algebra* it follows that A has exactly n eigenvalues counting with their multiplicities, i.e., (16.2) can be rewritten as

$$p(\lambda) = (-1)^n (\lambda - \lambda_1)^{r_1}(\lambda - \lambda_2)^{r_2} \cdots (\lambda - \lambda_m)^{r_m} = 0, \qquad (16.3)$$

where $\lambda_i \neq \lambda_j$, $r_i \geq 1$, $r_1 + \cdots + r_m = n$. The positive number r_i is called the *algebraic multiplicity* of the eigenvalue λ_i.

Knowing the distinct eigenvalues λ_i, $i = 1, \cdots, m$ $(\leq n)$ from (16.3), the corresponding eigenvectors can be obtained by computing nontrivial solutions of the homogenous systems $(A - \lambda_i I)u^i = 0$, $i = 1, \cdots, m$.

Example 16.1. The characteristic polynomial for the matrix

$$A = \begin{pmatrix} 2 & 1 & 0 \\ 1 & 3 & 1 \\ 0 & 1 & 2 \end{pmatrix} \qquad (16.4)$$

can be written as

$$p(\lambda) = -\lambda^3 + 7\lambda^2 - 14\lambda + 8 \; = \; -(\lambda - 1)(\lambda - 2)(\lambda - 4) \; = \; 0.$$

Thus, the eigenvalues are $\lambda_1 = 1$, $\lambda_2 = 2$, and $\lambda_3 = 4$. To find the corresponding eigenvectors, we need to find nonzero solutions of the systems $(A - \lambda_i I)u^i = 0$, $i = 1, 2, 3$. For $\lambda_1 = 1$, we need to solve

$$(A - \lambda_1 I)u^1 = \begin{pmatrix} 1 & 1 & 0 \\ 1 & 2 & 1 \\ 0 & 1 & 1 \end{pmatrix} \begin{pmatrix} u_1^1 \\ u_2^1 \\ u_3^1 \end{pmatrix} = \begin{matrix} u_1^1 + u_2^1 & = & 0 \\ u_1^1 + 2u_2^1 + u_3^1 & = & 0 \\ u_2^1 + u_3^1 & = & 0, \end{matrix}$$

which is the same as

$$\begin{matrix} u_1^1 + u_2^1 & = & 0 \\ u_2^1 + u_3^1 & = & 0 \\ u_2^1 + u_3^1 & = & 0. \end{matrix}$$

In this system the last two equations are the same, and we can take $u_3^1 = 1$ so that $u_2^1 = -1$, then the first equation gives $u_1^1 = 1$. Thus, $u^1 = (1, -1, 1)^t$. Similarly, we find $u^2 = (1, 0, -1)^t$ and $u^3 = (1, 2, 1)^t$.

In the case when the eigenvalues $\lambda_1, \cdots, \lambda_n$ of A are distinct it is easy to find the corresponding eigenvectors $u^1, \cdots, u^n$. For this, first we note that for the fixed eigenvalue λ_j of A at least one of the cofactors of $(a_{ii} - \lambda_j)$ in the matrix $(A - \lambda_j I)$ is nonzero. If not, then from (3.6) it follows that $p'(\lambda) = -[\text{cofactor of } (a_{11} - \lambda)] - \cdots - [\text{cofactor of } (a_{nn} - \lambda)]$, and hence $p'(\lambda_j) = 0$, i.e., λ_j was a multiple root, which is a contradiction to our assumption that λ_j is simple. Now let the cofactor of $(a_{kk} - \lambda_j)$ be different from zero, then one of the possible nonzero solutions of the system $(A - \lambda_j I)u^j = 0$ is $u_i^j = \text{cofactor of } a_{ki}$ in $(A - \lambda_j I)$, $1 \le i \le n$, $i \ne k$, $u_k^j = \text{cofactor of } (a_{kk} - \lambda_j)$ in $(A - \lambda_j I)$. Since for this choice of u^j, it follows from (3.2) that every equation, except the kth one, of the system $(A - \lambda_j I)u^j = 0$ is satisfied, and for the kth equation from (3.1), we have

$$\sum_{\substack{i=1 \\ i \ne k}}^n a_{ki}[\text{cofactor of } a_{ki}] + (a_{kk} - \lambda_j)[\text{cofactor of } (a_{kk} - \lambda_j)] \; = \; \det(A - \lambda_j I),$$

which is also zero. In conclusion this u^j is the eigenvector corresponding to the eigenvalue λ_j.

Example 16.2. Consider again the matrix A given in (16.4). Since in $(A - \lambda_1 I)$ the cofactor of $(a_{11} - \lambda_1) = 1 \ne 0$, we can take $u_1^1 = 1$, and then $u_2^1 = \text{cofactor of } a_{12} = -1$, $u_3^1 = \text{cofactor of } a_{13} = 1$, i.e., $u^1 = (1, -1, 1)^t$. Next, for $\lambda_2 = 2$ we have

$$(A - \lambda_2 I) = \begin{pmatrix} 0 & 1 & 0 \\ 1 & 1 & 1 \\ 0 & 1 & 0 \end{pmatrix}.$$

Since the cofactor of $(a_{22} - \lambda_2) = 0$ the choice $u_2^2 = $ cofactor of $(a_{22} - \lambda_2)$ is not correct. However, cofactor of $(a_{11} - \lambda_2) = $ cofactor of $(a_{33} - \lambda_2) = -1 \neq 0$ and we can take $u_1^2 = -1$ $(u_3^2 = -1)$, then $u_2^2 = $ cofactor of $a_{12} = 0$, $u_3^2 = $ cofactor of $a_{13} = 1$ $(u_1^2 = $ cofactor of $a_{31} = 1$, $u_2^2 = $ cofactor of $a_{32} = 0)$, i.e., $u^2 = (-1, 0, 1)^t$ $((1, 0, -1)^t)$. Similarly, we can find $u^3 = (1, 2, 1)^t$.

For the eigenvalues and eigenvectors of a given $n \times n$ matrix A the following properties are fundamental:

P1. There exists at least one eigenvector u associated with each distinct eigenvalue λ, and if A and λ are real, then u can be taken as real: $\det(A - \lambda I) = 0$ implies that the linear homogeneous system $(A - \lambda I)u = 0$ has at least one nontrivial solution. If A and λ are real, and the eigenvector $u = v + iw$, then $A(v + iw) = \lambda(v + iw)$, i.e., $Av = \lambda v$ and $Aw = \lambda w$. Now recall that both v and w are nonzero.

P2. If u is an eigenvector corresponding to the eigenvalue λ, then so is cu for all scalars $c \neq 0$: If $Au = \lambda u$, then $A(cu) = \lambda(cu)$.

P3. Let U_λ be the set of all eigenvectors associated with a given eigenvalue λ. Then, $V_\lambda = U_\lambda \cup \{0\} \subseteq R^n$ is an *invariant subspace* of A, i.e., $Au \in V_\lambda$ whenever $u \in V_\lambda$: Clearly, $0 \in V_\lambda$, and if $u, v \in V_\lambda$ and α, β are scalars, then $A(\alpha u + \beta v) = \alpha(Au) + \beta(Av) = \alpha(\lambda u) + \beta(\lambda v) = \lambda(\alpha u + \beta v)$. The space V_λ is called the *eigenspace* of A belonging to λ.

P4. If $\lambda_1, \cdots, \lambda_n$ are distinct eigenvalues of A and $u^1, \cdots, u^n$ are the corresponding eigenvectors, then the set $S = \{u^1, \cdots, u^n\}$ is linearly independent and forms a basis of R^n (C^n): If S is linearly dependent, then in view of Theorem 8.1 part 6, there exists the first u^r that is the linear combination of the preceding $u^1, \cdots, u^{r-1}$, i.e., there exist scalars $c_1, \cdots, c_{r-1}$ such that $u^r = c_1 u^1 + \cdots + c_{r-1} u^{r-1}$. Multiplying this relation by λ_r, we obtain

$$\lambda_r u^r = c_1 \lambda_r u^1 + \cdots + c_{r-1} \lambda_r u^{r-1},$$

whereas multiplying the same relation by A and using $Au^i = \lambda_i u^i$, we have

$$\lambda_r u^r = c_1 \lambda_1 u^1 + \cdots + c_{r-1} \lambda_{r-1} u^{r-1}.$$

Subtracting these equations, we find

$$0 = c_1(\lambda_r - \lambda_1)u^1 + \cdots + c_{r-1}(\lambda_r - \lambda_{r-1})u^{r-1};$$

however, since $u^1, \cdots, u^{r-1}$ are linearly independent and $\lambda_r \neq \lambda_i$, $i = 1, \cdots, r - 1$, it follows that $c_1 = \cdots = c_{r-1} = 0$. But then $u^r = 0$, which is impossible. The set S forms a basis that follows from Theorem 9.4.

P5. If $\lambda_1, \cdots, \lambda_m$ are distinct eigenvalues of A, and u^i is an eigenvector corresponding to the eigenvalue λ_i, then the set $S = \{u^1, \cdots, u^m\}$ is linearly independent.

P6. If (λ, u) is an eigenpair of A and A^{-1} exists, then $(1/\lambda, u)$ is an eigenpair of A^{-1}: If $Au = \lambda u$, then $A^{-1}(Au) = A^{-1}(\lambda u)$, i.e., $(1/\lambda)u = A^{-1}u$.

P7. The eigenvalues of A and A^t are the same; however, in general the eigenvectors are different: It follows from the fact that $\det(A - \lambda I) = \det(A^t - \lambda I)$. Now let A be real and have the distinct eigenvalues, and assume that the eigenvectors correspond to $\lambda_1, \cdots, \lambda_n$, $u^1, \cdots, u^n$, i.e., $Au^i = \lambda_i u^i$, $i = 1, \cdots, n$. Further, let the eigenvectors of A^t correspond to $\bar{\lambda}_1, \cdots, \bar{\lambda}_n$, $v^1, \cdots, v^n$, i.e., $A^t v^j = \bar{\lambda}_j v^j$, $j = 1, \cdots, n$. It follows that $(v^j)^H Au^i = \lambda_i (v^j)^H u^i$ and $(v^j)^H Au^i = \lambda_j (v^j)^H u^i$, and hence $(\lambda_i - \lambda_j)(v^j)^H u^i = 0$. Thus, for $i \neq j$, $(v^j)^H u^i = (v^j, u^i) = \overline{(u^i, v^j)} = 0$. Now since $\{v^i\}$ form a basis of R^n (C^n) we can decompose u^i as

$$u^i = c_1 v^1 + \cdots + c_i v^i + \cdots + c_n v^n,$$

which gives

$$0 < (u^i, u^i) = (u^i)^H u^i = \sum_{k=1}^{n} c_k (u^i)^H v^k = \sum_{k=1}^{n} c_k (u^i, v^k) = c_i (u^i, v^i),$$

and hence $(u^i)^H v^i \neq 0$. We can normalize the vectors so that $(u^i)^H v^i = 1$. In conclusion, we have

$$(v^j)^H u^i = \begin{cases} 0, & j \neq i \\ 1, & j = i, \end{cases} \tag{16.5}$$

and because of this property these sets of vectors $\{u^i\}$ and $\{v^i\}$ are called *biorthonormal*.

P8. The eigenvalues of a hermitian matrix are real, whereas those of a skew-hermitian matrix are purely imaginary. In particular, the eigenvalues of a real symmetric matrix are real, whereas of a real skew-symmetric matrix are purely imaginary: When $A = A^H$, $Au = \lambda u$ is the same as $u^H A = \bar{\lambda} u^H$. Thus, it follows that $u^H Au = \lambda u^H u$ and $u^H Au = \bar{\lambda} u^H u$, and hence $(\lambda - \bar{\lambda}) u^H u = 0$, but since $u^H u \neq 0$, we have $\lambda - \bar{\lambda} = 0$, and therefore λ is real. Similarly, for a skew-symmetric matrix it follows that $\lambda + \bar{\lambda} = 0$, and hence λ is complex.

P9. The eigenvectors of a real symmetric matrix that correspond to distinct eigenvalues are orthogonal: When $A^t = A$, $Au = \lambda u$ is the same as $u^t A = \lambda u^t$. Thus, if $Av = \mu v$, then since $u^t Av = \lambda u^t v$ it follows that $\mu u^t v = \lambda u^t v$, i.e., $(\mu - \lambda) u^t v = 0$, and therefore $u^t v = (u, v) = 0$.

P10. Let P be a nonsingular matrix. If (λ, u) is an eigenpair of A, then $(\lambda, P^{-1}u)$ is an eigenpair of $P^{-1}AP$: Clearly, $\det(A - \lambda I) = \det(P^{-1}(A - \lambda I)P) = \det(P^{-1}AP - \lambda P^{-1}P) = \det(P^{-1}AP - \lambda I)$. Further, since $Au = \lambda u$, we have $P^{-1}Au = \lambda P^{-1}u$, which is the same as $(P^{-1}AP)P^{-1}u = \lambda P^{-1}u$.

P11. If (λ, u) is an eigenpair of A, then (λ^m, u) is an eigenpair of A^m: It follows from $A^m u = A^{m-1}(Au) = A^{m-1}(\lambda u) = \lambda A^{m-1}u$. Thus, if $Q_m(x)$ is a polynomial, then $Q_m(\lambda)$ is an eigenvalue of $Q_m(A)$. In particular, the matrix A satisfies its own characteristic equation, i.e.,

$$p(A) = (-1)^n A^n + a_1 A^{n-1} + \cdots + a_{n-1}A + a_n = 0. \tag{16.6}$$

This result is known as the *Cayley–Hamilton* theorem.

Example 16.3. Multiplying (16.6) by A^{-1}, we find

$$A^{-1} = -\frac{1}{a_n}\left[(-1)^n A^{n-1} + a_1 A^{n-2} + \cdots + a_{n-1}I\right]. \tag{16.7}$$

Thus, if A^i, $i = 2, 3, \cdots, n-1$ are known then the inverse of the matrix A can be computed. In particular, for the matrix A in (16.4), we have

$$A^{-1} = -\frac{1}{8}\left[-A^2 + 7A - 14I\right]$$

$$= -\frac{1}{8}\left[-\begin{pmatrix} 5 & 5 & 1 \\ 5 & 11 & 5 \\ 1 & 5 & 5 \end{pmatrix} + 7\begin{pmatrix} 2 & 1 & 0 \\ 1 & 3 & 1 \\ 0 & 1 & 2 \end{pmatrix} - 14\begin{pmatrix} 1 & 0 & 0 \\ 0 & 1 & 0 \\ 0 & 0 & 1 \end{pmatrix}\right]$$

$$= \begin{pmatrix} \frac{5}{8} & -\frac{2}{8} & \frac{1}{8} \\ -\frac{2}{8} & \frac{4}{8} & -\frac{2}{8} \\ \frac{1}{8} & -\frac{2}{8} & \frac{5}{8} \end{pmatrix}.$$

P12. The number of linearly independent eigenvectors corresponding to an eigenvalue is called its *geometric multiplicity*. The geometric multiplicity of an eigenvalue is at most its algebraic multiplicity. Similar matrices A and B have the same eigenvalues, and these eigenvalues have the same algebraic and geometric multiplicities: Since there exists a nonsingular matrix P such that $B = P^{-1}AP$, we have $(B - \lambda I) = P^{-1}AP - \lambda P^{-1}P = P^{-1}(A - \lambda I)P$, i.e., if A and B are similar, then $(A - \lambda I)$ and $(B - \lambda I)$ are also similar. Thus from Problem 13.10 it follows that $\det(A - \lambda I) = \det(B - \lambda I)$ and $\mathcal{N}(A - \lambda I) = \mathcal{N}(B - \lambda I)$, i.e., A and B have the same eigenvalues with the same algebraic and geometric multiplicities.

Example 16.4. The eigenvalues of the matrix

$$A = \begin{pmatrix} 2 & 1 & -1 \\ -3 & -1 & 1 \\ 9 & 3 & -4 \end{pmatrix}$$

are $-1, -1, -1$. Further, the only linearly independent eigenvector corresponding to this eigenvalue is $(1, 0, 3)^t$. Hence the algebraic multiplicity of the eigenvalue -1 is 3, whereas its geometric multiplicity is one.

Example 16.5. The eigenvalues of the matrix

$$A = \begin{pmatrix} -1 & 0 & 4 \\ 0 & -1 & 2 \\ 0 & 0 & 1 \end{pmatrix}$$

are $-1, -1, 1$. Further, linearly independent eigenvectors corresponding to the eigenvalue -1 are $(1,0,0)^t$ and $(0,1,0)^t$. Hence the algebraic and geometric multiplicities of the eigenvalue -1 are 2. The eigenvector corresponding to the eigenvalue 1 is $(2,1,1)^t$.

The method discussed in this chapter to compute eigenvalues and eigenvectors is practical only for small matrices.

Problems

16.1. Let $\lambda_1, \cdots, \lambda_n$ be the (not necessarily distinct) eigenvalues of an $n \times n$ matrix A. Show that

(i)   for any constant α the eigenvalues of αA are $\alpha \lambda_1, \cdots, \alpha \lambda_n$

(ii) $\sum_{i=1}^{n} \lambda_i = \text{Tr } A$

(iii) $\prod_{i=1}^{n} \lambda_i = \det A$

(iv) if A is upper (lower) triangular, i.e., $a_{ij} = 0$, $i > j$ ($i < j$), then the eigenvalues of A are the diagonal elements of A

(v)   if A is real and λ_1 is complex with the corresponding eigenvector v^1, then there exists at least one i, $2 \le i \le n$, such that $\lambda_i = \bar{\lambda}_1$ and for such an i, $\bar{v}^1$ is the corresponding eigenvector

(vi) if $A^k = 0$ for some positive integer k, i.e., A is nilpotent, then 0 is the only eigenvalue of A

(vii) if A is orthogonal, then $|\lambda_i| = 1$, $i = 1, \cdots, n$.

16.2. Find the eigenvalues and eigenvectors of the following matrices

(i) $\begin{pmatrix} 4 & -2 & 1 \\ 1 & 3 & 0 \\ 2 & 8 & -1 \end{pmatrix}$ (ii) $\begin{pmatrix} 3 & 0 & 0 \\ -4 & 6 & 2 \\ 16 & -15 & -5 \end{pmatrix}$ (iii) $\begin{pmatrix} 8 & -6 & 2 \\ -6 & 7 & -4 \\ 2 & -4 & 3 \end{pmatrix}$.

16.3. Find the eigenvalues and eigenvectors of the following matrices

(i) $\begin{pmatrix} 4 & -2 \\ 5 & 2 \end{pmatrix}$ (ii) $\begin{pmatrix} 1 & 0 & 0 \\ 2 & 1 & -2 \\ 3 & 2 & 1 \end{pmatrix}$ (iii) $\begin{pmatrix} 2 & 1 & -2 \\ -1 & 0 & 0 \\ 1 & 1 & -1 \end{pmatrix}$.

16.4. Verify the Cayley–Hamilton theorem for the matrices given in Problem 16.2.

16.5. Use (16.7) to find the inverse of the matrices given in Problem 16.2.

16.6. Find algebraic and geometric multiplicities of each of the eigenvalues of the following matrices:

(i) $\begin{pmatrix} -1 & 1 & 0 \\ 0 & -1 & 0 \\ 0 & 0 & 3 \end{pmatrix}$ (ii) $\begin{pmatrix} 5 & -3 & -2 \\ 8 & -5 & -4 \\ -4 & 3 & 3 \end{pmatrix}$ (iii) $\begin{pmatrix} 2 & 1 & 1 \\ 0 & 2 & 0 \\ 0 & 0 & 3 \end{pmatrix}$.

16.7. The $n \times n$ matrix

$$P = \begin{pmatrix} 0 & 1 & 0 & \cdots & 0 \\ 0 & 0 & 1 & \cdots & 0 \\ \cdots & & & & \\ 0 & 0 & 0 & \cdots & 1 \\ -a_n & -a_{n-1} & -a_{n-2} & \cdots & -a_1 \end{pmatrix}$$

is called the *companion matrix*. Show that its characteristic polynomial is $p(\lambda) = (-1)^n(\lambda^n + a_1\lambda^{n-1} + \cdots + a_{n-1}\lambda + a_n)$, and corresponding to the eigenvalue λ, the eigenvector is $(1, \lambda, \lambda^2, \cdots, \lambda^{n-1})^t$.

16.8. A *circulant matrix* of order n is the $n \times n$ matrix defined by

$$C = \text{circ}(a_1, a_2, \cdots, a_n) = \begin{pmatrix} a_1 & a_2 & a_3 & \cdots & a_n \\ a_n & a_1 & a_2 & \cdots & a_{n-1} \\ a_{n-1} & a_n & a_1 & \cdots & a_{n-2} \\ \cdots & & & & \\ a_2 & a_3 & a_4 & \cdots & a_1 \end{pmatrix},$$

i.e., the elements of C in each row are the same as those in the previous row shifted one position to the right and wrapped around. Show that

(i) $C = a_1 I + a_2 P + \cdots + a_n P^{n-1}$, where P is the $n \times n$ companion matrix with $a_1 = \cdots = a_{n-1} = 0$, $a_n = -1$

(ii)   the eigenvalues λ_i and eigenvectors u^i, $i = 1, \cdots, n$ of C are $\lambda_i = a_1 + a_2\omega_i + \cdots + a_n\omega_i^{n-1}$ and $(1, \omega_i, \omega_i^2, \cdots, \omega_i^{n-1})^t$, where ω_i, $i = 1, \cdots, n$ are the nth roots of the unity, i.e., $\omega_i^n = 1$.

16.9. For the matrix $A_n(x)$ defined in (4.2), verify that the eigenvalues are

$$\lambda_i = x - 2\cos\frac{i\pi}{n+1}, \quad i = 1, \cdots, n.$$

In particular, for $x = 2$ the eigenvalues are

$$\lambda_i = 4\sin^2\frac{i\pi}{2(n+1)}, \quad i = 1, \cdots, n$$

and the eigenvector corresponding to λ_i is

$$\left(\sin\frac{i\pi}{n+1}, \sin\frac{2i\pi}{n+1}, \cdots, \sin\frac{ni\pi}{n+1} \right)^t.$$

16.10. Find the eigenvalues of the tridiagonal matrix $A = (a_{ij}) \in R^{n \times n}$, where

$$a_{ij} = \begin{cases} a, & i = j \\ b, & j - i = 1 \\ c, & i - j = 1 \\ 0, & |i - j| > 1 \end{cases}$$

and $bc > 0$.

16.11. Show that

$$\begin{vmatrix} x & 1 & & & & 1 \\ 1 & x & 1 & & & \\ & 1 & x & 1 & & \\ & & \cdots & \cdots & & \\ & & & 1 & x & 1 \\ 1 & & & & 1 & x \end{vmatrix} = \prod_{i=1}^{n} \left[x + 2 \cos \frac{2\pi i}{n} \right].$$

16.12. Show that

$$\begin{vmatrix} x & 2a & 1 & & & & & 1 & 2a \\ 2a & x & 2a & 1 & & & & & 1 \\ 1 & 2a & x & 2a & 1 & & & & \\ & 1 & 2a & x & 2a & 1 & & & \\ & & \cdots & \cdots & \cdots & \cdots & & & \\ & & & 1 & 2a & x & 2a & 1 & \\ 1 & & & & 1 & 2a & x & 2a \\ 2a & 1 & & & & 1 & 2a & x \end{vmatrix}$$

$$= \prod_{i=1}^{n} \left[x + 2 \cos \frac{4\pi i}{n} + 4a \cos \frac{2\pi i}{n} \right].$$

16.13. Let (V, F) be an n-dimensional vector space, and let $L : V \to V$ be a linear mapping. A nonzero vector $u \in V$ is called an *eigenvector* of L if there exists a $\lambda \in F$ such that $Au = \lambda u$. Show that

(i)   if $\lambda_1, \cdots, \lambda_n$ are distinct eigenvalues of L and $u^1, \cdots, u^n$ are the corresponding eigenvectors, then the set $S = \{u^1, \cdots, u^n\}$ is linearly independent, and forms a basis of V

(ii) L is one-to-one if and only if 0 is not an eigenvalue of L.

16.14. Find the eigenvalues and eigenvectors of $L : V \to V$, where L is defined by

(i) $V = R^4$, $L(x_1, x_2, x_3, x_4)^t = (x_1, x_1 + 5x_2 - 10x_3, x_1 + 2x_3, x_1 + 3x_4)^t$

(ii) V the space of twice continuously differential functions satisfying $x(0) = x(1) = 0$, $L(x) = -x''$

(iii) $V = (\mathcal{P}_2, R)$, $L(a + bx) = (a + b) + 2bx$.

16.15. Let A be an $n \times n$ matrix with linearly independent eigenvectors $u^1, \cdots, u^n$ and associated eigenvalues $\lambda_1, \cdots, \lambda_n$, where $|\lambda_1| > |\lambda_2| \geq |\lambda_3| \geq \cdots \geq |\lambda_n|$, i.e., λ_1 is the *dominant eigenvalue*). Further, let $\hat{u} = c_1 u^1 + \cdots + c_n u^n$, where $c_1 \neq 0$. Show that $\lim_{m \to \infty} (1/\lambda_1^m) A^m \hat{u} = c_1 u^1$. In particular, for the matrix A in (16.4) show that

$$\frac{1}{4^8} A^8 \begin{pmatrix} 1 \\ 1 \\ 1 \end{pmatrix} = \frac{1}{4^8} \begin{pmatrix} 11051 & 21845 & 10795 \\ 21845 & 43691 & 21845 \\ 10795 & 21845 & 11051 \end{pmatrix} \begin{pmatrix} 1 \\ 1 \\ 1 \end{pmatrix} \simeq \begin{pmatrix} 0.666672 \\ 1.333328 \\ 0.666672 \end{pmatrix} \simeq k \begin{pmatrix} 1 \\ 2 \\ 1 \end{pmatrix},$$

where k is a constant.

Answers or Hints

16.1. (i) $(\alpha A)x = (\alpha \lambda)x$.

(ii) $\det(A - \lambda I) = (a_{11} - \lambda) \cdot \text{cofactor}\,(a_{11} - \lambda) + \sum_{j=2}^{n} a_{1j} \cdot \text{cofactor}\, a_{1j}$, and since each term $a_{1j} \cdot \text{cofactor}\, a_{1j}$ is a polynomial of degree at most $n - 2$, on comparing the coefficients of λ^{n-1}, we get
$(-1)^{n+1} \sum_{i=1}^{n} \lambda_i = $ coefficient of λ^{n-1} in $(a_{11} - \lambda) \cdot \text{cofactor}\,(a_{11} - \lambda)$.
Therefore, an easy induction implies
$$\begin{aligned} (-1)^{n+1} \sum_{i=1}^{n} \lambda_i &= \text{coefficient of } \lambda^{n-1} \text{ in } (a_{11} - \lambda) \cdots (a_{nn} - \lambda) \\ &= (-1)^{n-1} \sum_{i=1}^{n} a_{ii}. \end{aligned}$$
(iii) In $\det(A - \lambda I) = (-1)^n (\lambda - \lambda_1) \cdots (\lambda - \lambda_n)$ substitute $\lambda = 0$.
(iv) Clear from the expansion of the determinant.
(v) For a polynomial with real coefficients, complex roots occur only in conjugate pairs. Thus $\lambda_1 = \mu + i\nu$ and $\overline{\lambda}_1 = \mu - i\nu$ both are eigenvalues. Now if $u = v + iw$ is an eigenvector corresponding to λ_1, i.e., $Au = \lambda_1 u$, then since A is real, $A\overline{u} = \overline{\lambda}_1 \overline{u}$.
(vi) Use P11.
(vii) If $A^{-1} = A^t$, then from P6 and P7, $1/\lambda = \lambda$, i.e., $\lambda^2 = 1$.

16.2. (i) $4, (1,1,2)^t; 3, (0,1,2)^t; -1, (-4,1,22)^t$

(ii) $0, (0,1,-3)^t; 1, (0,-2,5)^t; 3, (1,0,2)^t$

(iii) $0, (1,2,2)^t; 3, (2,1,-2)^t; 15, (2,-2,1)^t$.

16.3. (i) $3 \pm 3i, (2,1 \mp 3i)^t$.

(ii) $1, (-2,3,-2)^t; 1 \pm 2i, (0,\pm i,1)^t$.

(iii) $1, (1,-1,0)^t; \pm i, (1,\pm i,1)$.

16.4. Verify directly.

16.5. (i) $\frac{1}{12} \begin{pmatrix} 3 & -6 & 3 \\ -1 & 6 & -1 \\ -2 & 36 & -14 \end{pmatrix}$.

(ii) Singular matrix.

(iii) Singular matrix.

16.6. (i) For $\lambda_1 = -1$, a.m. is 2 and g.m. is 1. For $\lambda_2 = 3$, a.m. and g.m. are 1.

(ii)   For $\lambda = 1$, a.m. is 3 and g.m. is 2.

(iii)   For $\lambda_1 = 2$, a.m. is 2 and g.m. is 1. For $\lambda_2 = 3$, a.m. and g.m. are 1.

16.7. $C_1 + \lambda C_2 + \cdots + \lambda^{n-1} C_n$.

16.8. (i) Verify directly.

(ii)   The characteristic polynomial of P is $(-1)^n(\lambda^n - 1)$. Use property P12.

16.9. Use (4.6) with x replaced by $x - \lambda$. Verify directly.

16.10. $\lambda_i = a - 2\sqrt{bc}\cos\frac{i\pi}{n+1}$, $i = 1, \cdots, n$.

16.11. The corresponding matrix is circulant with $a_1 = x, a_2 = 1, a_3 = \cdots = a_{n-1} = 0, a_n = 1$. Thus from Problem 16.7, its eigenvalues are $\lambda_k = x + \omega_k + \omega_k^{n-1}$. Now, since $\omega_k = e^{2k\pi i/n}$, it follows that $\lambda_k = x + 2\cos(2\pi k/n)$. Now use Problem 16.1(iii).

16.12. The corresponding matrix is circulant with eigenvalues $\lambda_k = x + 2a\omega_k + \omega_k^2 + \omega_k^{n-2} + 2a\omega_k^{n-1}$.

16.13. (i) Similar to P4.

(ii)   Similar to Theorem 12.3.

16.14. (i)   The eigenvalues and the eigenvectors are $1, 2, 3, 5$ and $(4, -11, -4, 0)^t, (0, 10, 3, 0)^t, (0, 0, 0, 1)^t, (0, 1, 0, 0)^t$.

(ii) $-x'' = \lambda x$, $x(0) = x(1) = 0$, $\lambda_n = n^2\pi^2$, $x_n(t) = \sin n\pi t$, $n = 1, 2, \cdots$.

(iii)   The eigenvalues and the eigenvectors are $1, 2$ and $(1, 0)^t, (1, 1)^t$.

16.15. $A^m\hat{u} = A^m(c_1 u^1 + \cdots + c_n u^n) = c_1\lambda_1^m u^1 + \cdots + c_n\lambda_n^m u^n$, and hence
$$\frac{1}{\lambda_1^m} A^m\hat{u} = c_1 u^1 + \left(\frac{\lambda_2}{\lambda_1}\right)^m u^2 + \cdots + \left(\frac{\lambda_n}{\lambda_1}\right)^m u^n.$$

Chapter 17

Normed Linear Spaces

The distance between a vector and the zero vector is a measure of the length of the vector. This generalized notion, which plays a central role in the whole of analysis, is called a norm. We begin this chapter with the definition of a norm of a vector and then extend it to a matrix. Then we will drive some estimates on the eigenvalues of a given matrix. Some very useful convergence results are also proved.

A *norm* (or *length*) on a vector space V is a function that assigns to each vector $u \in V$ a nonnegative real number, denoted as $\|u\|$, which satisfies the following axioms:

1. *Positive definite property:* $\|u\| \geq 0$, and $\|u\| = 0$ if and only if $u = 0$,
2. *Homogeneity property:* $\|cu\| = |c|\|u\|$ for each scalar c,
3. *Triangle inequality:* $\|u + v\| \leq \|u\| + \|v\|$ for all $u, v \in V$.

A vector space V with a norm $\| \cdot \|$ is called a *normed linear space*, and is denoted as $(V, \| \cdot \|)$. In the vector space R^n (C^n) the following three norms are in common use:

$$\text{absolute norm } \|u\|_1 = \sum_{i=1}^{n} |u_i|,$$

$$\text{Euclidean norm } \|u\|_2 = \left(\sum_{i=1}^{n} |u_i|^2 \right)^{1/2} = \sqrt{(u, u)},$$

and

$$\text{maximum norm } \|u\|_\infty = \max_{1 \leq i \leq n} |u_i|.$$

As an example, for the vector $u = (1, 2, -3, 5)^t$, we have $\|u\|_1 = 11$, $\|u\|_2 = \sqrt{39}$, $\|u\|_\infty = 5$, whereas for the vector $u = (1 + i, 2 - 3i, 4)$, $\|u\|_1 = \sqrt{2} + \sqrt{13} + 4$, $\|u\|_2 = \sqrt{31}$, $\|u\|_\infty = 4$.

The notations $\| \cdot \|_1$, $\| \cdot \|_2$, and $\| \cdot \|_\infty$ are justified because of the fact that all these norms are special cases of the general ℓ_p–norm

$$\|u\|_p = \left(\sum_{i=1}^{n} |u_i|^p \right)^{1/p}, \quad p \geq 1.$$

Similarly, in the vector space of real-valued continuous functions $C[a, b]$, the following three norms are frequently used:

$$\|f\|_1 = \int_a^b |f(x)| dx, \quad \|f\|_2 = \left(\int_a^b f^2(x) dx \right)^{1/2} \quad \text{and} \quad \|f\|_\infty = \max_{x \in [a,b]} |f(x)|.$$

Theorem 17.1 (Cauchy–Schwarz inequality).

For any pair of vectors u, v in an inner product space V, the following inequality holds:

$$|(u, v)| \leq \|u\|_2 \|v\|_2. \tag{17.1}$$

Proof. If $v = 0$, there is nothing to prove. If $v \neq 0$, for $\lambda \in R$, we have

$$
\begin{aligned}
0 &\leq \|u - (u, v)\lambda v\|_2^2 \\
&= (u - (u, v)\lambda v, u - (u, v)\lambda v) \\
&= (u, u) - \overline{(u, v)}\lambda(u, v) - (u, v)\lambda(v, u) + (u, v)\overline{(u, v)}\lambda^2(v, v) \\
&= \|u\|_2^2 - 2\lambda|(u, v)|^2 + |(u, v)|^2 \lambda^2 \|v\|_2^2.
\end{aligned}
$$

Now let $\lambda = 1/\|v\|_2^2$, to obtain

$$0 \leq \|u\|_2^2 - \frac{|(u, v)|^2}{\|v\|_2^2},$$

which immediately gives (17.1). ∎

Using (17.1) in (14.1) it follows that $-1 \leq \cos\theta \leq 1$, and hence the angle θ between two vectors in a real inner product space (V, R) exists and is unique. Thus, in (V, R), $|(u, v)| = \|u\|_2 \|v\|_2$ provided $\theta = 0$, i.e., u and v are orthogonal.

Theorem 17.2 (Minkowski inequality).

For any pair of vectors u, v in an inner product space V, the following inequality holds:

$$\|u + v\|_2 \leq \|u\|_2 + \|v\|_2. \tag{17.2}$$

Proof. In view of the inequality (17.1) and the properties of inner products, we have

$$
\begin{aligned}
\|u + v\|_2^2 = (u + v, u + v) &= (u, u) + (u, v) + (v, u) + (v, v) \\
&= \|u\|_2^2 + (u, v) + \overline{(u, v)} + \|v\|_2^2 \\
&= \|u\|_2^2 + 2\text{Re}\,(u, v) + \|v\|^2 \\
&\leq \|u\|_2^2 + 2\|u\|_2\|v\|_2 + \|v\|_2^2 \\
&= (\|u\|_2 + \|v\|_2)^2,
\end{aligned}
$$

which on taking the square root of both sides yields (17.2). ∎

Example 17.1. For the vectors $u = (a_1, \cdots, a_n), v = (b_1, \cdots, n_n) \in C^n$, inequality (17.1) reduces to

$$(a_1\bar{b}_1 + \cdots + a_n\bar{b}_n) \leq (|a_1|^2 + \cdots + |a_n|^2)^{1/2}(|b_1|^2 + \cdots + |b_n|^2)^{1/2}. \quad (17.3)$$

Example 17.2. For the functions $f, g \in C[a, b]$, inequality (17.1) reduces to

$$\int_a^b f(x)\bar{g}(x)dx \leq \left(\int_a^b |f(x)|^2 dx\right)^{1/2} \left(\int_a^b |g(x)|^2\right)^{1/2}. \quad (17.4)$$

The set $C^{n \times n}$ can be considered as equivalent to the vector space C^{n^2}, with a special multiplicative operation added into the vector space. Thus, a matrix norm should satisfy the usual three requirements of a vector norm and, in addition, we require:

4. $\|AB\| \leq \|A\|\|B\|$ for all $n \times n$ matrices A, B (and hence, in particular, for every positive integer p, $\|A^p\| \leq \|A\|^p$),

5. compatibility with the vector norm, i.e., if $\|\cdot\|_*$ is the norm in C^n, then $\|Au\|_* \leq \|A\|\|u\|_*$ for all $u \in C^n$ and any $n \times n$ matrix A.

Once a norm $\|\cdot\|_*$ is fixed, in C^n then an associated matrix norm is usually defined by

$$\|A\| = \sup_{u \neq 0} \frac{\|Au\|_*}{\|u\|_*} = \sup_{\|u\|_* = 1} \|Au\|_*. \quad (17.5)$$

From (17.5) condition 5 is immediately satisfied. To show 4 we use 5 twice, to obtain

$$\|ABu\|_* = \|A(Bu)\|_* \leq \|A\|\|Bu\|_* \leq \|A\|\|B\|\|u\|_*$$

and hence for all $u \neq 0$, we have

$$\frac{\|ABu\|_*}{\|u\|_*} \leq \|A\|\|B\|,$$

or

$$\|AB\| = \sup_{u \neq 0} \frac{\|ABu\|_*}{\|u\|_*} \leq \|A\|\|B\|.$$

The norm of the matrix A induced by the vector norm $\|u\|_*$ will be denoted by $\|A\|_*$. For the three norms $\|u\|_1$, $\|u\|_2$, and $\|u\|_\infty$, the corresponding matrix norms are

$$\|A\|_1 = \max_{1 \leq j \leq n} \sum_{i=1}^n |a_{ij}|, \quad \|A\|_2 = \sqrt{\rho(A^t A)}, \quad \text{and} \quad \|A\|_\infty = \max_{1 \leq i \leq n} \sum_{j=1}^n |a_{ij}|,$$

where for a given $n \times n$ matrix A with eigenvalues $\lambda_1, \cdots, \lambda_n$ not necessarily distinct, $\rho(A)$ is called the *spectral radius* of A and is defined as

$$\rho(A) = \max\{|\lambda_i|, \ 1 \leq i \leq n\}.$$

Theorem 17.3. For a given square matrix A, $\rho(A) \leq \|A\|$.

Proof. Since $Ax = \lambda x$, we have $\|\lambda x\| = \|Ax\| \leq \|A\|\|x\|$, which is the same as $|\lambda|\|x\| \leq \|A\|\|x\|$, and hence $|\lambda| \leq \|A\|$. ∎

From Theorem 17.3 it follows that all eigenvalues of A lie in or on the disk $|z| \leq \|A\|$ of the complex plane C; in particular, in the disks

$$|z| \leq \max_{1 \leq i \leq n} \sum_{j=1}^{n} |a_{ij}| \tag{17.6}$$

and

$$|z| \leq \max_{1 \leq j \leq n} \sum_{i=1}^{n} |a_{ij}|. \tag{17.7}$$

Remark 17.1. Let (λ, u) be an eigenpair of A, and let $|u_i| = \max\{|u_1|, \cdots, |u_n|\}$. In component form, $Au = \lambda u$ can be written as

$$(\lambda - a_{ii})u_i = \sum_{j=1, j \neq i}^{n} a_{ij}u_j, \quad i = 1, \cdots, n,$$

and hence it follows that

$$|\lambda - a_{ii}| \leq \sum_{j=1, j \neq i}^{n} |a_{ij}|.$$

Therefore, all the eigenvalues of A lie inside or on the union of the circles

$$|z - a_{ii}| \leq \sum_{j=1, j \neq i}^{n} |a_{ij}|, \quad i = 1, \cdots, n. \tag{17.8}$$

Also, since the eigenvalues of A and A^t are the same, all the eigenvalues of A lie inside or on the union of the circles

$$|z - a_{jj}| \leq \sum_{i=1, i \neq j}^{n} |a_{ij}|, \quad j = 1, \cdots, n. \tag{17.9}$$

The above estimates are attributed to Gershgorin. Since all these bounds are independent, all the eigenvalues of A must lie in the intersection of these circles. If any one of these circles is isolated, then it contains exactly one eigenvalue.

Example 17.3. For the matrix A in (16.4), in view of (17.6) as well as (17.7), all the eigenvalues lie in or on the circle $|z| \leq 5$; whereas, in view of (17.8) as well as (17.9), all the eigenvalues lie inside or on the union of the circles $|z - 2| \leq 1$ and $|z - 3| \leq 2$, which is $|z - 3| \leq 2$. Now, since the intersection of $|z| \leq 5$ and $|z - 3| \leq 2$ is $|z - 3| \leq 2$, all the eigenvalues of A lie in or on the circle $|z - 3| \leq 2$.

Remark 17.2. If all the elements of A are positive, then the numerically largest eigenvalue is also positive, and the corresponding eigenvector has positive coordinates. This result is due to Perron. For the matrix A in (16.4), the largest eigenvalue is 4 and the corresponding eigenvector is $(1, 2, 1)^t$.

Remark 17.3. Let A be a hermitian matrix with (real) eigenvalues $\lambda_1, \cdots, \lambda_n$. Define $\lambda = \min\{\lambda_1, \cdots, \lambda_n\}$ and $\Lambda = \max\{\lambda_1, \cdots, \lambda_n\}$. Then, for any nonzero vector $u \in C^n$, the following inequality holds:

$$\lambda \leq \frac{(Au, u)}{(u, u)} \leq \Lambda. \tag{17.10}$$

Further, $\lambda = \min_{\|u\|=1}(Au, u)$ and $\Lambda = \max_{\|u\|=1}(Au, u)$. The expression $(Au, u)/(u, u)$ is called the *Rayleigh quotient*.

A sequence of vectors $\{u^m\}$ in a normed linear space V is said to converge to $u \in V$ if and only if $\|u - u^m\| \to 0$ as $m \to \infty$. A sequence of matrices $\{A^{(m)}\}$ with elements $(a_{ij}^{(m)})$ is said to converge to $A = (a_{ij})$ if and only if $a_{ij} = \lim_{m \to \infty} a_{ij}^{(m)}$. Further, the series $\sum_{m=1}^{\infty} A^{(m)}$ is convergent if and only if the sequence of its partial sums, i.e., $\{\sum_{m=1}^{k} A^{(m)}\}$, converges.

In a normed linear space V, norms $\|\cdot\|$ and $\|\cdot\|_*$ are said to be *equivalent* if there exist positive constants m and M such that for all $u \in V$, $m\|u\| \leq \|u\|_* \leq M\|u\|$. It is well known (see Problem 17.7) that in a finite dimensional normed linear space all the norms are equivalent. Thus, if a sequence $\{u^m\}$ converges in one norm then it converges in all the norms.

Theorem 17.4. For a given square matrix A, $\lim_{m \to \infty} A^m = 0$ if $\|A\| < 1$.

Proof. Since $\|A^m\| \leq \|A\|^m$ and the norm is continuous (see Problem 17.2), it follows that $\|\lim_{m \to \infty} A^m\| \leq \lim_{m \to \infty} \|A\|^m = 0$. ∎

Theorem 17.5. For a given square matrix A, $(I - A)^{-1}$ exists if $\|A\| < 1$. Further,

$$I + A + A^2 + \cdots = \sum_{m=0}^{\infty} A^m = (I - A)^{-1}. \tag{17.11}$$

Proof. Since

$$\|(I - A)x\| = \|x - Ax\| \geq \|x\| - \|Ax\| \geq \|x\| - \|A\|\|x\| = (1 - \|A\|)\|x\|,$$

$(I - A)x \neq 0$ for an arbitrary $x \neq 0$, which implies that $(I - A)$ is a nonsingular matrix. Now, after multiplying the identity

$$(I + A + A^2 + \cdots + A^m)(I - A) \;=\; I - A^{m+1},$$

by $(I - A)^{-1}$, we obtain

$$(I + A + A^2 + \cdots + A^m) \;=\; (I - A^{m+1})(I - A)^{-1}.$$

In the above equality as $m \to \infty$, (17.11) follows. ∎

Theorem 17.6 (Banach's lemma). For a given square matrix A, if $\|A\| < 1$, then

$$\frac{1}{1 + \|A\|} \;\leq\; \|(I \pm A)^{-1}\| \;\leq\; \frac{1}{1 - \|A\|}. \tag{17.12}$$

Proof. Since

$$I \;=\; (I - A)(I - A)^{-1} \;=\; (I - A)^{-1} - A(I - A)^{-1},$$

we have

$$\|A\|\,\|(I - A)^{-1}\| \;\geq\; \|A(I - A)^{-1}\| \;=\; \|(I - A)^{-1} - I\| \;\geq\; \|(I - A)^{-1}\| - 1,$$

and hence $\|(I-A)^{-1}\| \leq 1/(1-\|A\|)$. Analogously, letting $I = (I+A)(I+A)^{-1}$ we find $\|(I + A)^{-1}\| \geq 1/(1 + \|A\|)$. Finally, since $\| - A\| = \|A\|$, inequalities (17.12) follow. ∎

Problems

17.1. Show that for all vectors $u, v \in R^3$,

(i)   *Lagrange's identity* $\|u \times v\|_2^2 = \|u\|_2^2\|v\|_2^2 - (u \cdot v)^2$

(ii) $\|u \times v\|_2 = \|u\|_2\|v\|_2 \sin\theta$, where θ is the angle between u and v.

17.2. Show that
$$\big|\,\|u\| - \|v\|\,\big| \;\leq\; \|u - v\|.$$

Thus the norm is a *Lipschitz function* and, therefore, in particular, a continuous real valued function.

17.3. For any pair of vectors u, v in an inner product space V, show that

(i) $\|u + v\|_2^2 + \|u - v\|_2^2 = 2\|u\|_2^2 + 2\|v\|_2^2$ (parallelogram law)

(ii) $\mathrm{Re}\,(u, v) = \dfrac{1}{4}\|u + v\|_2^2 - \dfrac{1}{4}\|u - v\|_2^2.$

In particular, for the vectors $u = (2, 0, 1, 3)^t$, $v = (3, 2, 1, 0)^t$ in R^4, verify the above relations.

17.4. Let $\{u^1, \cdots, u^r\}$ be an orthogonal subset of an inner product space V. Show that the *generalized theorem of Pythagoras*, i.e.,

$$\|u^1 + \cdots + u^r\|_2^2 = \|u^1\|_2^2 + \cdots + \|u^r\|_2^2$$

holds. In particular, verify this theorem for the orthogonal set $\{(0,1,1)^t, (1,-\frac{1}{2},\frac{1}{2})^t, (\frac{2}{3},\frac{2}{3},-\frac{2}{3})^t\}$ obtained in Example 14.7.

17.5. Let $S = \{u^1, \cdots, u^r\}$ be an orthogonal subset of an inner product space V. Show that for any vector $v \in V$, the following holds:

$$\left\| v - \sum_{i=1}^r c_i u^i \right\|_2 \leq \left\| v - \sum_{i=1}^r d_i u^i \right\|_2 ,$$

where $c_i = (u^i, v)/(u^i, u^i)$, $i = 1, \cdots, r$ are the Fourier coefficients, and d_i, $i = 1, \cdots, r$ are arbitrary scalars. Thus, in ℓ_2-norm, $\sum_{i=1}^r c_i u^i = \sum_{i=1}^r \text{proj}_{u^i} v = \text{proj}_S v$ is the closest (*best approximation*) to v as a linear combination of $u^1, \cdots, u^r$. Thus, in view of Example 14.7 from the vector $(2,2,3)^t$ to the set $\text{Span}\{(0,1,1)^t, (1-1/2,1/2)^t\}$, the minimum ℓ_2-distance is

$$\left\| (2,2,3)^t - \frac{5}{2}(0,1,1)^t - \frac{5}{3}\left(1,-\frac{1}{2},\frac{1}{2}\right)^t \right\|_2 = \frac{1}{\sqrt{3}}.$$

17.6. Let $\{\tilde{u}^1, \cdots, \tilde{u}^r\}$ be an orthonormal subset of an inner product space V. Show that for any vector $v \in V$, Bessel's inequality holds:

$$\sum_{i=1}^r |c_i|^2 \leq \|v\|_2^2,$$

where $c_i = (v, \tilde{u}^i)$, $i = 1, \cdots, r$ are the Fourier coefficients. Verify this inequality for the orthonormal set $\left\{ \frac{1}{\sqrt{2}}(0,1,1)^t, \sqrt{\frac{2}{3}}(1,-\frac{1}{2},\frac{1}{2})^t, \right\}$ and the vector $(2,2,3)^t$.

17.7. Let $q \geq p \geq 1$. Show that

(i) for any $x \in \mathbb{R}^n$,

$$\|x\|_q \leq \|x\|_p \leq n^{(q-p)/pq}\|x\|_q$$

(ii) for any $n \times n$ matrix A,

$$n^{(p-q)/pq}\|A\|_q \leq \|A\|_p \leq n^{(q-p)/pq}\|A\|_q.$$

17.8. Let A be an $n \times n$ real matrix. Show that $A^t A$ has nonnegative eigenvalues.

17.9. Let A and B be $n \times n$ matrices. The matrix A is nonsingular and $\|A^{-1}B\| < 1$. Show that $A + B$ is nonsingular, and

$$\|(A+B)^{-1} - A^{-1}\| \leq \frac{\|A^{-1}B\|}{1 - \|A^{-1}B\|}\|A^{-1}\|.$$

17.10. Let V be a normed linear space. The *distance function* between two vectors $u, v \in V$ is defined by $d(u,v) = \|u - v\|$. Show that

(i) $d(u,v) \geq 0$, and $d(u,v) = 0$ if and only if $u = v$

(ii) $d(u,v) = d(v,u)$

(iii) $d(u,v) \leq d(u,w) + d(w,v)$ for every $w \in V$.

For $d(u,v) = \|u - v\|_2$ the above definition reduces to the familiar Euclidean distance. In particular, for the vectors $u = (1,2,3)^t$, $v = (2,0,1)^t$, $w = (1,3,0)^t$ in $(R^3, \|\cdot\|_2)$, verify the above properties.

Answers or Hints

17.1. (i) $\|u \times v\|_2^2 = \|(u_2v_3 - u_3v_2, u_3v_1 - u_1v_3, u_1v_2 - u_2v_1)\|_2^2 = |u_2v_3 - u_3v_2|^2 + |u_3v_1 - u_1v_3|^2 + |u_1v_2 - u_2v_1|^2 = \|u\|_2^2\|v\|_2^2 - (u,v)^2$.

(ii) $\|u \times v\|_2^2 = \|u\|_2^2\|v\|_2^2 - (\|u\|_2\|v\|_2 \cos\theta)^2$.

17.2. $\|u\| = \|u - v + v\| \leq \|u - v\| + \|v\|$.

17.3. (i) $\|u+v\|_2^2 + \|u-v\|_2^2 = (u+v, u+v) + (u-v, u-v) = 2(u,u) + 2(v,v)$.

(ii) $\frac{1}{4}\|u + v\|_2^2 - \frac{1}{4}\|u - v\|_2^2 = \frac{1}{2}[(u,v) + (v,u)]$.

17.4. $\|\sum_{i=1}^{n} u^i\|_2^2 = (\sum_{i=1}^{n} u^i, \sum_{i=1}^{n} u^i) = \sum_{i=1}^{n}(u^i, u^i) + \sum_{i \neq j}(u^i, u^j)$.

17.5. By Theorem 14.3, $v - \sum_{i=1}^{r} c_i u^i$ is orthogonal to every u^i, and hence orthogonal to any linear combination of $u^1, \cdots, u^r$. Thus, from Problem 17.5 it follows that

$\|v - \sum_{i=1}^{r} d_i u^i\|_2^2 = \|(v - \sum_{i=1}^{r} c_i u^i) + (\sum_{i=1}^{r}(c_i - d_i)u^i)\|_2^2$
$= \|v - \sum_{i=1}^{r} c_i u^i\|_2^2 + \|\sum_{i=1}^{r}(c_i - d_i)u^i\|_2^2 \geq \|v - \sum_{i=1}^{r} c_i u^i\|_2^2$.

17.6. $0 \leq (v - \sum_{i=1}^{r} c_i \tilde{u}^i, v - \sum_{i=1}^{r} c_i \tilde{u}^i) = \|v\|^2 - 2\text{Re}\,(v, \sum_{i=1}^{r} c_i \tilde{u}^i) + \sum_{i=1}^{r} |c_i|^2 = \|v\|^2 - \sum_{i=1}^{r} |c_i|^2$.

17.7. (i) First we will show that for $0 < p < q$, $\|x\|_p \geq \|x\|_q$, $x \in R^n$. If $x = 0$, then it is obviously true. Otherwise, let $y_k = |x_k|/\|x\|_q$. Clearly, $y_k \leq 1$ for all $k = 1, \cdots, n$. Therefore, $y_k^p \geq y_k^q$, and hence $\|y\|_p \geq 1$, which implies $\|x\|_p \geq \|x\|_q$. To prove the right side of the inequality, we need Hölder's inequality

$$\sum_{i=1}^{n} |u_i v_i| \leq \left(\sum_{i=1}^{n} |u_i|^r\right)^{1/r} \left(\sum_{i=1}^{n} |v_i|^s\right)^{1/s}, \quad r > 1, \; \frac{1}{r} + \frac{1}{s} = 1.$$

In this inequality, we let $u_i = |x_i|^p$, $v_i = 1$, $r = q/p > 1$, $s = q/(q-p)$, to get

$$\sum_{i=1}^{n} |x_i|^p \leq \left(\sum_{i=1}^{n} |x_i|^q\right)^{p/q} (n)^{(q-p)/q}.$$

(ii) For $q \geq p \geq 1$, from (i), we have

$$\|A\|_p = \max_{x \neq 0} \frac{\|Ax\|_p}{\|x\|_p} \leq \max_{x \neq 0} \frac{\|Ax\|_p}{\|x\|_q} \leq \max_{x \neq 0} \frac{n^{\frac{q-p}{pq}}\|Ax\|_q}{\|x\|_q} = n^{\frac{q-p}{pq}}\|A\|_q,$$

$$\|A\|_p = \max_{x \neq 0} \frac{\|Ax\|_p}{\|x\|_p} \geq \max_{x \neq 0} \frac{\|Ax\|_p}{n^{\frac{q-p}{pq}}\|x\|_q} \geq \max_{x \neq 0} \frac{\|Ax\|_q}{n^{\frac{q-p}{pq}}\|x\|_q} = n^{\frac{p-q}{pq}}\|A\|_q.$$

17.8. Since $A^t A$ is real and symmetric, in view of P8 (in Chapter 16), eigenvalues of $A^t A$ are real. If λ is an eigenvalue of $A^t A$ and u is the corresponding eigenvector, then $\|Au\|_2^2 = ((Au),(Au)) = (Au)^t(Au) = u^t(A^t A)u = u^t \lambda u = \lambda u^t u = \lambda \|u\|_2^2$, and hence λ is nonnegative.

17.9. Since $\|A^{-1}B\| < 1$ from Theorem 17.5, it follows that the matrix $(I + A^{-1}B)$ is nonsingular. Now since $A + B = A(I + A^{-1}B)$, the matrix $A + B$ is nonsingular, and $(A+B)^{-1} - A^{-1} = (I + A^{-1}B)^{-1}A^{-1} - A^{-1} = [(I + A^{-1}B)^{-1} - I]A^{-1}$. Now use (17.12).

17.10. (i) If $u \neq v$, then $u - v \neq 0$. Hence $d(u,v) = \|u - v\| > 0$. Further, $d(u,u) = \|u - u\| = \|0\| = 0$

(ii) $d(u,v) = \|u - v\| = \| -1(v - u)\| = |-1|\|v - u\| = \|v - u\|$

(iii) $d(u,v) = \|u - v\| = \|(u - w) + (w - v)\| \leq \|u - w\| + \|w - v\| = d(u,w) + d(w,v)$.

Chapter 18

Diagonalization

An $n \times n$ matrix A is said to be *diagonalizable* if there exists a nonsingular matrix P and a diagonal matrix D such that $A = PDP^{-1}$, which is the same as $D = P^{-1}AP$. An immediate advantage of diagonalization is that we can find powers of A immediately. In fact, note that

$$A^2 = (PDP^{-1})(PDP^{-1}) = PD(P^{-1}P)DP^{-1} = PD^2P^{-1},$$

and for any positive integer m, it follows that

$$A^m = PD^mP^{-1}.$$

From Chapter 2, we also recall that if the diagonal elements of D are $(\lambda_1, \cdots, \lambda_n)$, then D^m is also diagonal, with diagonal elements $(\lambda_1^m, \cdots, \lambda_n^m)$.

Our main result of this chapter is the following theorem.

Theorem 18.1. An $n \times n$ matrix A is diagonalizable if and only if A has n linearly independent eigenvectors, i.e., algebraic multiplicity of each eigenvalue is the same as the geometric multiplicity. Further, if $D = P^{-1}AP$, where D is a diagonal matrix, then the diagonal elements of D are the eigenvalues of A and the column vectors of P are the corresponding eigenvectors.

Proof. Let $\lambda_1, \cdots, \lambda_n$ be the eigenvalues (not necessarily distinct) of A and let $u^1, \cdots, u^n$ be the corresponding linearly independent eigenvectors. We define an $n \times n$ matrix P whose i-th column is the vector u^i. Clearly, in view of Problem 8.3 the matrix P is invertible. Now, since

$$Au^i = \lambda_i u^i,$$

it follows that

$$AP = A(u^1, \cdots, u^n) = (Au^1, \cdots, Au^n) = (\lambda_1 u^1, \cdots, \lambda_n u^n) = PD,$$

where D is the diagonal matrix with diagonal elements $\lambda_1, \cdots, \lambda_n$. Thus, $AP = PD$, and hence $D = P^{-1}AP$.

Conversely, suppose that A is diagonalizable, i.e., there exist a diagonal matrix D and an invertible matrix P such that $D = P^{-1}AP$. Again assume that the diagonal elements of D are $(\lambda_1, \cdots, \lambda_n)$ and the column vectors of P

are $(u^1, \cdots, u^n)$. Since $AP = PD$, it follows that $Au^i = \lambda_i u^i$, $i = 1, \cdots, n$. Hence, $u^1, \cdots, u^n$ are eigenvectors of A. Since P is invertible, from Problem 8.3 it follows that $u^1, \cdots, u^n$ are linearly independent. ∎

Theorem 18.1 says that the matrices A and D are similar, and hence in view of P12 (in Chapter 16) both have the same eigenvalues.

Corollary 18.1. If A is an $n \times n$ matrix with n distinct eigenvalues, then A is diagonalizable.

Proof. It follows from P4 (in Chapter 16) and Theorem 18.1. ∎

Example 18.1. In view of Example 16.1 for the matrix A in (16.4), we have

$$D = \begin{pmatrix} 1 & 0 & 0 \\ 0 & 2 & 0 \\ 0 & 0 & 4 \end{pmatrix} \quad \text{and} \quad P = \begin{pmatrix} 1 & 1 & 1 \\ -1 & 0 & 2 \\ 1 & -1 & 1 \end{pmatrix}.$$

Now, since

$$P^{-1} = \frac{1}{6} \begin{pmatrix} 2 & -2 & 2 \\ 3 & 0 & -3 \\ 1 & 2 & 1 \end{pmatrix}$$

from Theorem 18.1, it follows that

$$\begin{pmatrix} 1 & 0 & 0 \\ 0 & 2 & 0 \\ 0 & 0 & 4 \end{pmatrix} = \frac{1}{6} \begin{pmatrix} 2 & -2 & 2 \\ 3 & 0 & -3 \\ 1 & 2 & 1 \end{pmatrix} \begin{pmatrix} 2 & 1 & 0 \\ 1 & 3 & 1 \\ 0 & 1 & 2 \end{pmatrix} \begin{pmatrix} 1 & 1 & 1 \\ -1 & 0 & 2 \\ 1 & -1 & 1 \end{pmatrix}$$

and

$$A = \begin{pmatrix} 2 & 1 & 0 \\ 1 & 3 & 1 \\ 0 & 1 & 2 \end{pmatrix} = \begin{pmatrix} 1 & 1 & 1 \\ -1 & 0 & 2 \\ 1 & -1 & 1 \end{pmatrix} \begin{pmatrix} 1 & 0 & 0 \\ 0 & 2 & 0 \\ 0 & 0 & 4 \end{pmatrix} \frac{1}{6} \begin{pmatrix} 2 & -2 & 2 \\ 3 & 0 & -3 \\ 1 & 2 & 1 \end{pmatrix}.$$

For each positive integer m we also have

$$A^m = \begin{pmatrix} 1 & 1 & 1 \\ -1 & 0 & 2 \\ 1 & -1 & 1 \end{pmatrix} \begin{pmatrix} 1 & 0 & 0 \\ 0 & 2^m & 0 \\ 0 & 0 & 4^m \end{pmatrix} \frac{1}{6} \begin{pmatrix} 2 & -2 & 2 \\ 3 & 0 & -3 \\ 1 & 2 & 1 \end{pmatrix}$$

$$= \begin{pmatrix} \frac{2}{6} + \frac{3}{6}2^m + \frac{1}{6}4^m & -\frac{2}{6} + \frac{2}{6}4^m & \frac{2}{6} - \frac{3}{6}2^m + \frac{1}{6}4^m \\ -\frac{2}{6} + \frac{2}{6}4^m & \frac{2}{6} + \frac{4}{6}4^m & -\frac{2}{6} + \frac{2}{6}4^m \\ \frac{2}{6} - \frac{3}{6}2^m + \frac{1}{6}4^m & -\frac{2}{6} + \frac{2}{6}4^m & \frac{2}{6} + \frac{3}{6}2^m + \frac{1}{6}4^m \end{pmatrix}.$$

Example 18.2. For the matrix A in Example 16.5, we find

$$D = \begin{pmatrix} -1 & 0 & 0 \\ 0 & -1 & 0 \\ 0 & 0 & 1 \end{pmatrix} \quad \text{and} \quad P = \begin{pmatrix} 1 & 0 & 2 \\ 0 & 1 & 1 \\ 0 & 0 & 1 \end{pmatrix}.$$

Now, since

$$P^{-1} = \begin{pmatrix} 1 & 0 & -2 \\ 0 & 1 & -1 \\ 0 & 0 & 1 \end{pmatrix}$$

from Theorem 18.1, it follows that

$$A = \begin{pmatrix} -1 & 0 & 4 \\ 0 & -1 & 2 \\ 0 & 0 & 1 \end{pmatrix} = \begin{pmatrix} 1 & 0 & 2 \\ 0 & 1 & 1 \\ 0 & 0 & 1 \end{pmatrix} \begin{pmatrix} -1 & 0 & 0 \\ 0 & -1 & 0 \\ 0 & 0 & 1 \end{pmatrix} \begin{pmatrix} 1 & 0 & -2 \\ 0 & 1 & -1 \\ 0 & 0 & 1 \end{pmatrix}.$$

Example 18.3. In view of Theorem 18.1, the matrix A in Example 16.4 cannot be diagonalized.

Remark 18.1. The matrix P that diagonalizes the matrix A is not unique. For example, if in Example 18.1, we take

$$D = \begin{pmatrix} 2 & 0 & 0 \\ 0 & 4 & 0 \\ 0 & 0 & 1 \end{pmatrix} \quad \text{and} \quad P = \begin{pmatrix} 1 & 1 & 1 \\ 0 & 2 & -1 \\ -1 & 1 & 1 \end{pmatrix},$$

then

$$A = \begin{pmatrix} 2 & 1 & 0 \\ 1 & 3 & 1 \\ 0 & 1 & 2 \end{pmatrix} = \begin{pmatrix} 1 & 1 & 1 \\ 0 & 2 & -1 \\ -1 & 1 & 1 \end{pmatrix} \begin{pmatrix} 2 & 0 & 0 \\ 0 & 4 & 0 \\ 0 & 0 & 1 \end{pmatrix} \frac{1}{6} \begin{pmatrix} 3 & 0 & -3 \\ 1 & 2 & 1 \\ 2 & -2 & 2 \end{pmatrix}.$$

Similarly, in Example 18.2, we could have taken

$$D = \begin{pmatrix} -1 & 0 & 0 \\ 0 & 1 & 0 \\ 0 & 0 & -1 \end{pmatrix} \quad \text{and} \quad P = \begin{pmatrix} 1 & 2 & 0 \\ 0 & 1 & 1 \\ 0 & 1 & 0 \end{pmatrix}.$$

Then,

$$A = \begin{pmatrix} -1 & 0 & 4 \\ 0 & -1 & 2 \\ 0 & 0 & 1 \end{pmatrix} = \begin{pmatrix} 1 & 2 & 0 \\ 0 & 1 & 1 \\ 0 & 1 & 0 \end{pmatrix} \begin{pmatrix} -1 & 0 & 0 \\ 0 & 1 & 0 \\ 0 & 0 & -1 \end{pmatrix} \begin{pmatrix} 1 & 0 & -2 \\ 0 & 0 & 1 \\ 0 & 1 & -1 \end{pmatrix}.$$

A *linear mapping* $L : V \rightarrow V$ is called *diagonalizable* if there is a basis S for V such that the transition matrix A for L relative to S is a diagonalizable matrix.

Remark 18.2. In the above definition, let S and T be two bases of V, and let A and B be the corresponding transition matrices. Then, in view of Chapter 13, A and B are similar, i.e., there exists an invertible matrix Q such that $B = QAQ^{-1}$. Thus, if A is diagonalizable, i.e., $A = PDP^{-1}$, then $B = QPDP^{-1}Q^{-1} = (QP)D(QP)^{-1}$, and hence B is diagonalizable. Therefore, in the above definition, if L is diagonalizable with respect to one basis, it is diagonalizable with respect to all bases. However, with respect to different bases, the corresponding transition matrices and their diagonalizations may be different (see Example 18.4 and Problem 18.2).

Example 18.4. Consider the linear mapping $L : R^3 \to R^3$ as $L(x, y, z) = (11x - y - 4z, -x + 11y - 4z, -4x - 4y + 14z)^t$ and the basis S_1 for R^3 as $S_1 = \{e^1, e^2, e^3\}$. For this mapping, the transition matrix A relative to S_1 is

$$A = \begin{pmatrix} 11 & -1 & -4 \\ -1 & 11 & -4 \\ -4 & -4 & 14 \end{pmatrix}. \tag{18.1}$$

For this matrix A the eigenvalues and the corresponding eigenvectors are

$$\begin{aligned} \lambda_1 &= 6, & u^1 &= (1, 1, 1)^t \\ \lambda_2 &= 12, & u^2 &= (-1, 1, 0)^t \\ \lambda_3 &= 18, & u^3 &= (-1, -1, 2)^t. \end{aligned}$$

Thus this mapping is diagonalizable. Further, it follows that

$$A = PDP^{-1} = \begin{pmatrix} 1 & -1 & -1 \\ 1 & 1 & -1 \\ 1 & 0 & 2 \end{pmatrix} \begin{pmatrix} 6 & 0 & 0 \\ 0 & 12 & 0 \\ 0 & 0 & 18 \end{pmatrix} \frac{1}{6} \begin{pmatrix} 2 & 2 & 2 \\ -3 & 3 & 0 \\ -1 & -1 & 2 \end{pmatrix}. \tag{18.2}$$

Remark 18.3. The matrix A in (18.1) is symmetric, its eigenvalues are real, and its eigenvectors are orthogonal, as they should be in view of P8 and P9 (in Chapter 16). The columns of P are orthogonal (but not the row vectors) and the row vectors of P^{-1} are also orthogonal (but not the column vectors). Clearly, we can normalize the above eigenvectors, and then in (18.2) the matrix P can be replaced by

$$Q = \begin{pmatrix} \frac{1}{\sqrt{3}} & -\frac{1}{\sqrt{2}} & -\frac{1}{\sqrt{6}} \\ \frac{1}{\sqrt{3}} & \frac{1}{\sqrt{2}} & -\frac{1}{\sqrt{6}} \\ \frac{1}{\sqrt{3}} & 0 & \frac{2}{\sqrt{6}} \end{pmatrix}.$$

This matrix Q is orthogonal (rows as well as columns are orthonormal, see Problem 14.11) and hence $Q^t = Q^{-1}$. Thus, it follows that

$$A = QDQ^t = \begin{pmatrix} \frac{1}{\sqrt{3}} & -\frac{1}{\sqrt{2}} & -\frac{1}{\sqrt{6}} \\ \frac{1}{\sqrt{3}} & \frac{1}{\sqrt{2}} & -\frac{1}{\sqrt{6}} \\ \frac{1}{\sqrt{3}} & 0 & \frac{2}{\sqrt{6}} \end{pmatrix} \begin{pmatrix} 6 & 0 & 0 \\ 0 & 12 & 0 \\ 0 & 0 & 18 \end{pmatrix} \begin{pmatrix} \frac{1}{\sqrt{3}} & \frac{1}{\sqrt{3}} & \frac{1}{\sqrt{3}} \\ -\frac{1}{\sqrt{2}} & \frac{1}{\sqrt{2}} & 0 \\ -\frac{1}{\sqrt{6}} & -\frac{1}{\sqrt{6}} & \frac{2}{\sqrt{6}} \end{pmatrix}.$$

An $n \times n$ matrix A is said to be *orthogonally diagonalizable* if there exists an orthogonal matrix Q and a diagonal matrix D such that $A = QDQ^{-1} = QDQ^t$. Thus the matrix A in (18.1), which is real and symmetric, is orthogonally diagonalizable. In fact, we have the following general result.

Theorem 18.2. A real $n \times n$ matrix A is orthogonally diagonalizable if and only if A is symmetric.

Proof. If $A = QDQ^t$, then

$$A^t = (QDQ^t)^t = (Q^t)^t D^t Q^t = QDQ^t = A,$$

i.e., $A^t = A$, and hence A is symmetric.

For the converse, we note the following facts: From P8 (in Chapter 16), eigenvalues of a real symmetric matrix are real; from P9 (in Chapter 16), eigenvectors of a real symmetric matrix that correspond to distinct eigenvalues are orthogonal; for a symmetric matrix, algebraic multiplicity and the geometric multiplicity of an eigenvalue are the same; however, if the geometric multiplicity of an eigenvalue is greater than 1, then the corresponding eigenvectors (though linearly independent) may not be mutually orthogonal, but the Gram–Schmidt process can be used to orthogonalize them. All of these vectors can be orthonormalized. ∎

Example 18.5. For the symmetric matrix

$$A = \begin{pmatrix} 1 & 1 & 1 & 1 \\ 1 & \frac{5}{3} & -\frac{4}{3} & -\frac{4}{3} \\ 1 & -\frac{4}{3} & -\frac{5}{6} & \frac{7}{6} \\ 1 & -\frac{4}{3} & \frac{7}{6} & -\frac{5}{6} \end{pmatrix},$$

the eigenvalues and the corresponding eigenvectors are

$$\begin{array}{llll} \lambda_1 &=& 2, & u^1 = (3,1,1,1)^t \\ \lambda_2 &=& 3, & u^2 = (0,-2,1,1)^t \\ \lambda_3 &=& -2, & u^3 = (-1,1,2,0)^t \\ \lambda_4 &=& -2, & u^4 = (0,0,-1,1). \end{array}$$

Clearly, the sets $S_1 = \{u^1, u^2, u^3\}$ and $S_2 = \{u^1, u^2, u^4\}$ are orthogonal; however, the vectors u^3 and u^4, although linearly independent, are not orthogonal. We use the Gram–Schmidt process to orthogonalize the vectors u^3 and u^4, to obtain $v^3 = (-1,1,2,0)^t$ and $v^4 = (-1/3,1/3,-1/3,1)^t$. The set $S = \{u^1, u^2, v^3, v^4\}$ is orthogonal. Next, we normalize these vectors, to find

$$Q = \begin{pmatrix} \frac{3}{\sqrt{12}} & 0 & -\frac{1}{\sqrt{6}} & -\frac{1}{\sqrt{12}} \\ \frac{1}{\sqrt{12}} & -\frac{2}{\sqrt{6}} & \frac{1}{\sqrt{6}} & \frac{1}{\sqrt{12}} \\ \frac{1}{\sqrt{12}} & \frac{1}{\sqrt{6}} & \frac{2}{\sqrt{6}} & -\frac{1}{\sqrt{12}} \\ \frac{1}{\sqrt{12}} & \frac{1}{\sqrt{6}} & 0 & \frac{3}{\sqrt{12}} \end{pmatrix}.$$

We further note that in the factorization QDQ^t of the given matrix A, the matrix D is

$$D = \begin{pmatrix} 2 & 0 & 0 & 0 \\ 0 & 3 & 0 & 0 \\ 0 & 0 & -2 & 0 \\ 0 & 0 & 0 & -2 \end{pmatrix}.$$

Finally, in this chapter we prove the following theorem, known as QR *factorization.*

Theorem 18.3. Let $A = (a^1, \cdots, a^n)$ be an $m \times n$ matrix with linearly independent columns. Then, A can be factorized as $A = QR$, where $Q = (q^1, \cdots, q^n)$ is an $m \times n$ matrix with orthonormal columns and R is an $n \times n$ upper triangular matrix with positive diagonal elements.

Proof. To columns of A we apply the Gram–Schmidt process to obtain the required matrix Q with orthonormal columns $\{q^1, \cdots, q^n\}$. For each a^j, $1 \le j \le n$ by the Gram–Schmidt process ensures that $a^j \in \text{Span}\{q^1, \cdots, q^j\}$. Thus, from (14.5) it follows that

$$a^j = (a^j, q^1)q^1 + (a^j, q^2)q^2 + \cdots + (a^j, q^j)q^j. \tag{18.3}$$

Let $r_{kj} = (a^j, q^k)$, $1 \le j \le n, 1 \le k \le j$, and define the matrix

$$R = \begin{pmatrix} r_{11} & r_{12} & \cdots & r_{1n} \\ 0 & r_{22} & \cdots & r_{2n} \\ \vdots & \vdots & \vdots & \vdots \\ 0 & 0 & \cdots & r_{nn} \end{pmatrix}.$$

Now we claim that $A = QR$. For this, it suffices to note that the j-th column of QR is exactly (18.3). Now, clearly $a^k \notin \text{Span}\{q^1, \cdots, q^{k-1}\}$, and hence $r_{kk} = (a^k, q^k) \neq 0$. If $r_{kk} < 0$, without affecting the orthonormality of $\{q^1, \cdots, q^n\}$, we replace q^k with $-q^k$, which will make $r_{kk} > 0$. Hence, the diagonal entries of R can be made positive. ∎

Remark 18.4. Since $A = QR$, we have $Q^t A = Q^t QR = R$, and hence once Q is known, the matrix R can be computed immediately.

Example 18.6. From Problem 14.14(ii) it follows that for the matrix

$$A = (a^1, a^2, a^3) = \begin{pmatrix} 1 & 1 & 2 \\ 1 & 2 & 3 \\ 1 & 3 & 4 \\ 3 & 4 & 9 \end{pmatrix}$$

the matrix Q is

$$Q = (q^1, q^2, q^3) = \begin{pmatrix} \frac{1}{2\sqrt{3}} & -\frac{1}{2\sqrt{3}} & -\frac{2}{\sqrt{6}} \\ \frac{1}{2\sqrt{3}} & \frac{1}{2\sqrt{3}} & -\frac{1}{\sqrt{6}} \\ \frac{1}{2\sqrt{3}} & \frac{3}{2\sqrt{3}} & 0 \\ \frac{3}{2\sqrt{3}} & -\frac{1}{2\sqrt{3}} & \frac{1}{\sqrt{6}} \end{pmatrix}.$$

Now, directly or using Remark 18.4, we can compute the matrix R as

$$R = \begin{pmatrix} r_{11} & r_{12} & r_{13} \\ 0 & r_{22} & r_{23} \\ 0 & 0 & r_{33} \end{pmatrix} = \begin{pmatrix} (a^1, q^1) & (a^2, q^1) & (a^3, q^1) \\ 0 & (a^2, q^2) & (a^3, q^2) \\ 0 & 0 & (a^3, q^3) \end{pmatrix} = \begin{pmatrix} \frac{6}{\sqrt{3}} & \frac{9}{\sqrt{3}} & \frac{18}{\sqrt{3}} \\ 0 & \frac{3}{\sqrt{3}} & \frac{2}{\sqrt{3}} \\ 0 & 0 & \frac{2}{\sqrt{6}} \end{pmatrix}.$$

Problems

18.1. If possible, diagonalize matrices given in Problems 16.2, 16.3, and 16.6.

18.2. Diagonalize the following matrices

(i) $\begin{pmatrix} 10 & 4 & -2 \\ -10 & 24 & -2 \\ -20 & 8 & 16 \end{pmatrix}$, (ii) $\begin{pmatrix} 5 & -1 & -1 & -1 \\ -1 & 5 & 1 & 1 \\ -1 & 1 & 7 & -1 \\ -1 & 1 & -1 & 7 \end{pmatrix}.$

18.3. In Example 18.4 with the given basis S_i find the corresponding matrices P^i and D^i, $i = 2, \cdots, 7$

(i) $S_2 = \{e^2, e^3, e^1\}$, (ii) $S_3 = \{e^3, e^1, e^2\}$, (iii) $S_4 = \{e^1, e^3, e^2\}$, (iv) $S_5 = \{e^2, e^1, e^3\}$, (v) $S_6 = \{e^3, e^2, e^1\}$ (vi) $S_7 = \{e^1, e^1 + e^2, e^1 + e^2 + e^3\}$.

18.4. Orthogonally diagonalize the following matrices, and the matrix in Problem 18.2(ii):

(i) $\begin{pmatrix} 2 & 0 & 0 \\ 0 & \frac{3}{2} & -\frac{1}{2} \\ 0 & -\frac{1}{2} & \frac{3}{2} \end{pmatrix}$, (ii) $\begin{pmatrix} 13 & 2 & 4 \\ 2 & 10 & 2 \\ 4 & 2 & 13 \end{pmatrix}$, (iii) $\begin{pmatrix} \frac{3}{2} & 0 & \frac{1}{2} & 0 \\ 0 & \frac{10}{9} & 0 & \frac{4}{9\sqrt{2}} \\ \frac{1}{2} & 0 & \frac{3}{2} & 0 \\ 0 & \frac{4}{9\sqrt{2}} & 0 & \frac{17}{9} \end{pmatrix}.$

18.5. Find QR factorization of the following matrices:

(i) $\begin{pmatrix} 0 & 1 & 1 \\ 1 & 0 & 1 \\ 1 & 1 & 0 \end{pmatrix}$, (ii) $\begin{pmatrix} 1 & 0 \\ 1 & 2 \\ 1 & 0 \\ -1 & 3 \end{pmatrix}$, (iii) $\begin{pmatrix} 1 & 0 & -2 \\ 1 & 2 & -1 \\ 1 & 0 & 3 \\ -1 & 3 & 0 \end{pmatrix}$.

Answers or Hints

18.1. 16.2(i) $\begin{pmatrix} 1 & 0 & -4 \\ 1 & 1 & 1 \\ 2 & 2 & 22 \end{pmatrix} \begin{pmatrix} 4 & 0 & 0 \\ 0 & 3 & 0 \\ 0 & 0 & -1 \end{pmatrix} \begin{pmatrix} 1 & -\frac{2}{5} & \frac{1}{5} \\ -1 & \frac{3}{2} & -\frac{1}{4} \\ 0 & -\frac{1}{10} & \frac{1}{20} \end{pmatrix}$.

16.2(ii) $\begin{pmatrix} 0 & 0 & 1 \\ 1 & -2 & 0 \\ -3 & 5 & 2 \end{pmatrix} \begin{pmatrix} 0 & 0 & 0 \\ 0 & 1 & 0 \\ 0 & 0 & 3 \end{pmatrix} \begin{pmatrix} 4 & -5 & -2 \\ 2 & -3 & -1 \\ 1 & 0 & 0 \end{pmatrix}$.

16.2(iii) $\begin{pmatrix} 1 & 2 & 2 \\ 2 & 1 & -2 \\ 2 & -2 & 1 \end{pmatrix} \begin{pmatrix} 0 & 0 & 0 \\ 0 & 3 & 0 \\ 0 & 0 & 15 \end{pmatrix} \begin{pmatrix} \frac{1}{9} & \frac{2}{9} & \frac{2}{9} \\ \frac{2}{9} & \frac{1}{9} & -\frac{2}{9} \\ \frac{2}{9} & -\frac{2}{9} & \frac{1}{9} \end{pmatrix}$.

16.3(i) $\begin{pmatrix} 2 & 2 \\ 1-3i & 1+3i \end{pmatrix} \begin{pmatrix} 3+3i & 0 \\ 0 & 3-3i \end{pmatrix} \begin{pmatrix} \frac{1}{12}(3-i) & \frac{1}{6}i \\ \frac{1}{12}(3+i) & -\frac{1}{6}i \end{pmatrix}$.

16.3(ii) $\begin{pmatrix} -2 & 0 & 0 \\ 3 & i & -i \\ -2 & 1 & 1 \end{pmatrix} \begin{pmatrix} 1 & 0 & 0 \\ 0 & 1+2i & 0 \\ 0 & 0 & 1-2i \end{pmatrix} \begin{pmatrix} -\frac{1}{2} & 0 & 0 \\ -\frac{1}{2}-\frac{3}{4}i & -\frac{1}{2}i & \frac{1}{2} \\ -\frac{1}{2}+\frac{3}{4}i & \frac{1}{2}i & \frac{1}{2} \end{pmatrix}$.

16.3(iii) $\begin{pmatrix} 1 & 1 & 1 \\ -1 & i & -i \\ 0 & 1 & 1 \end{pmatrix} \begin{pmatrix} 1 & 0 & 0 \\ 0 & i & 0 \\ 0 & 0 & -i \end{pmatrix} \begin{pmatrix} 1 & 0 & -1 \\ -\frac{1}{2}i & -\frac{1}{2}i & \frac{1}{2}+\frac{1}{2}i \\ \frac{1}{2}i & \frac{1}{2}i & \frac{1}{2}-\frac{1}{2}i \end{pmatrix}$.

16.6(i), (ii), (iii) None of them are diagonalizable.

18.2. (i) $\begin{pmatrix} -1 & 2 & 1 \\ 0 & 5 & 1 \\ 5 & 0 & 2 \end{pmatrix} \begin{pmatrix} 20 & 0 & 0 \\ 0 & 20 & 0 \\ 0 & 0 & 10 \end{pmatrix} \begin{pmatrix} -\frac{2}{5} & \frac{4}{25} & \frac{3}{25} \\ -\frac{1}{5} & \frac{7}{25} & -\frac{1}{25} \\ 1 & -\frac{2}{5} & \frac{1}{5} \end{pmatrix}$.

(ii) $\begin{pmatrix} 1 & 2 & -1 & -1 \\ 1 & 0 & 1 & 1 \\ 0 & 1 & 2 & 0 \\ 0 & 1 & 0 & 2 \end{pmatrix} \begin{pmatrix} 4 & 0 & 0 & 0 \\ 0 & 4 & 0 & 0 \\ 0 & 0 & 8 & 0 \\ 0 & 0 & 0 & 8 \end{pmatrix} \frac{1}{8} \begin{pmatrix} 2 & 6 & -2 & -2 \\ 2 & -2 & 2 & 2 \\ -1 & 1 & 3 & -1 \\ -1 & 1 & -1 & 3 \end{pmatrix}$.

18.3. Most are approximate values

(i) $P^2 = \begin{pmatrix} 1 & -0.583333-0.702179i & -0.583333+0.702179i \\ 1 & -0.416667+0.702179i & -0.416667-0.702179i \\ 1 & 1 & 1 \end{pmatrix}$

$D^2 = \begin{pmatrix} 6 & 0 & 0 \\ 0 & \frac{3}{2}(-5+i\sqrt{71}) & 0 \\ 0 & 0 & -\frac{3}{2}(5+i\sqrt{71}) \end{pmatrix}$.

(ii) $P^3 = \begin{pmatrix} 1 & -0.416667 + 0.702179i & -0.416667 - 0.702179i \\ 1 & -0.583333 - 0.702179i & -0.583333 + 0.702179i \\ 1 & 1 & 1 \end{pmatrix}$

$D^3 = \begin{pmatrix} 6 & 0 & 0 \\ 0 & \frac{3}{2}(-5 + i\sqrt{71}) & 0 \\ 0 & 0 & -\frac{3}{2}(5 + i\sqrt{71}) \end{pmatrix}$.

(iii) $P^4 = \begin{pmatrix} 1 & -1.7374 & -0.0959285 \\ 1 & 0.737405 & -0.904071 \\ 1 & 1 & 1 \end{pmatrix}$

$D^4 = \begin{pmatrix} 6 & 0 & 0 \\ 0 & \frac{3}{2}(-1 + \sqrt{97}) & 0 \\ 0 & 0 & -\frac{3}{2}(1 + \sqrt{97}) \end{pmatrix}$.

(iv) $P^5 = \begin{pmatrix} 1 & -1 & -1 \\ 1 & 1 & -1 \\ 1 & 0 & 2 \end{pmatrix}$ $\quad D^5 = \begin{pmatrix} 6 & 0 & 0 \\ 0 & -12 & 0 \\ 0 & 0 & 18 \end{pmatrix}$.

(v) $P^6 = \begin{pmatrix} 0.603423 & -1.25509 & 0.76072 \\ -0.760381 & -0.168228 & 1.5603 \\ 1 & 1 & 1 \end{pmatrix}$

$D^6 = \begin{pmatrix} 15.4894 & 0 & 0 \\ 0 & -12.8983 & 0 \\ 0 & 0 & 8.4089 \end{pmatrix}$.

(vi) $P^7 = \begin{pmatrix} 2.50743 & -1.00372 - 0.581853i & -1.00372 + 0.581853i \\ -1.45332 & 0.0391602 - 0.80221i & 0.0391602 + 0.80221i \\ 1 & 1 & 1 \end{pmatrix}$

$D^7 = \begin{pmatrix} 7.59684 & 0 & 0 \\ 0 & 9.70158 + 8.74509i & 0 \\ 0 & 0 & 9.70158 - 8.74509i \end{pmatrix}$.

18.4. (i) $\begin{pmatrix} 0 & 1 & 0 \\ \frac{1}{\sqrt{2}} & 0 & -\frac{1}{\sqrt{2}} \\ \frac{1}{\sqrt{2}} & 0 & \frac{1}{\sqrt{2}} \end{pmatrix} \begin{pmatrix} 1 & 0 & 0 \\ 0 & 2 & 0 \\ 0 & 0 & 2 \end{pmatrix} \begin{pmatrix} 0 & \frac{1}{\sqrt{2}} & \frac{1}{\sqrt{2}} \\ 1 & 0 & 0 \\ 0 & -\frac{1}{\sqrt{2}} & \frac{1}{\sqrt{2}} \end{pmatrix}$, or

$\begin{pmatrix} 0 & \frac{2}{\sqrt{6}} & \frac{1}{\sqrt{3}} \\ \frac{1}{\sqrt{2}} & -\frac{1}{\sqrt{6}} & \frac{1}{\sqrt{3}} \\ \frac{1}{\sqrt{2}} & \frac{1}{\sqrt{6}} & -\frac{1}{\sqrt{3}} \end{pmatrix} \begin{pmatrix} 1 & 0 & 0 \\ 0 & 2 & 0 \\ 0 & 0 & 2 \end{pmatrix} \begin{pmatrix} 0 & \frac{1}{\sqrt{2}} & \frac{1}{\sqrt{2}} \\ \frac{2}{\sqrt{6}} & -\frac{1}{\sqrt{6}} & \frac{1}{\sqrt{6}} \\ \frac{1}{\sqrt{3}} & \frac{1}{\sqrt{3}} & -\frac{1}{\sqrt{3}} \end{pmatrix}$.

(ii) $\begin{pmatrix} -\frac{2}{3} & \frac{1}{3} & \frac{2}{3} \\ \frac{2}{3} & \frac{2}{3} & \frac{1}{3} \\ \frac{1}{3} & -\frac{2}{3} & \frac{2}{3} \end{pmatrix} \begin{pmatrix} 9 & 0 & 0 \\ 0 & 9 & 0 \\ 0 & 0 & 18 \end{pmatrix} \begin{pmatrix} -\frac{2}{3} & \frac{2}{3} & \frac{1}{3} \\ \frac{1}{3} & \frac{2}{3} & -\frac{2}{3} \\ \frac{2}{3} & \frac{1}{3} & \frac{2}{3} \end{pmatrix}$, or

$\begin{pmatrix} -\frac{1}{\sqrt{5}} & -\frac{4}{3\sqrt{5}} & \frac{2}{3} \\ \frac{2}{\sqrt{5}} & -\frac{2}{3\sqrt{5}} & \frac{1}{3} \\ 0 & \frac{5}{3\sqrt{5}} & \frac{2}{3} \end{pmatrix} \begin{pmatrix} 9 & 0 & 0 \\ 0 & 9 & 0 \\ 0 & 0 & 18 \end{pmatrix} \begin{pmatrix} -\frac{1}{\sqrt{5}} & \frac{2}{\sqrt{5}} & 0 \\ -\frac{4}{3\sqrt{5}} & -\frac{2}{3\sqrt{5}} & \frac{5}{3\sqrt{5}} \\ \frac{2}{3} & \frac{1}{3} & \frac{2}{3} \end{pmatrix}$.

(iii)
$$\begin{pmatrix} \frac{1}{2} & \frac{1}{2} & -\frac{1}{2} & -\frac{1}{2} \\ \frac{1}{3\sqrt{2}} & \frac{2}{3} & \frac{2}{3} & \frac{1}{3\sqrt{2}} \\ \frac{1}{2} & -\frac{1}{2} & \frac{1}{2} & -\frac{1}{2} \\ \frac{2}{3} & -\frac{1}{3\sqrt{2}} & -\frac{1}{3\sqrt{2}} & \frac{2}{3} \end{pmatrix} \begin{pmatrix} 2 & 0 & 0 & 0 \\ 0 & 1 & 0 & 0 \\ 0 & 0 & 1 & 0 \\ 0 & 0 & 0 & 2 \end{pmatrix} \begin{pmatrix} \frac{1}{2} & \frac{1}{3\sqrt{2}} & \frac{1}{2} & \frac{2}{3} \\ \frac{1}{2} & \frac{2}{3} & -\frac{1}{2} & -\frac{1}{3\sqrt{2}} \\ -\frac{1}{2} & \frac{2}{3} & \frac{1}{2} & -\frac{1}{3\sqrt{2}} \\ -\frac{1}{2} & \frac{1}{3\sqrt{2}} & -\frac{1}{2} & \frac{2}{3} \end{pmatrix}.$$

18.2(ii)
$$\begin{pmatrix} \frac{1}{\sqrt{2}} & \frac{1}{2} & -\frac{1}{\sqrt{6}} & \frac{1}{2\sqrt{3}} \\ \frac{1}{\sqrt{2}} & -\frac{1}{2} & \frac{1}{\sqrt{6}} & \frac{1}{2\sqrt{3}} \\ 0 & \frac{1}{2} & \frac{2}{\sqrt{6}} & -\frac{1}{2\sqrt{3}} \\ 0 & \frac{1}{2} & 0 & \frac{3}{2\sqrt{3}} \end{pmatrix} \begin{pmatrix} 4 & 0 & 0 & 0 \\ 0 & 4 & 0 & 0 \\ 0 & 0 & 8 & 0 \\ 0 & 0 & 0 & 8 \end{pmatrix} \begin{pmatrix} \frac{1}{\sqrt{2}} & \frac{1}{\sqrt{2}} & 0 & 0 \\ \frac{1}{2} & -\frac{1}{2} & \frac{1}{2} & \frac{1}{2} \\ -\frac{1}{\sqrt{6}} & \frac{1}{\sqrt{6}} & \frac{2}{\sqrt{6}} & 0 \\ -\frac{1}{2\sqrt{3}} & \frac{1}{2\sqrt{3}} & -\frac{1}{2\sqrt{3}} & \frac{3}{2\sqrt{3}} \end{pmatrix}.$$

18.5. (i)
$$\begin{pmatrix} 0 & \sqrt{\frac{2}{3}} & \frac{1}{\sqrt{3}} \\ \frac{1}{\sqrt{2}} & -\frac{1}{2}\sqrt{\frac{2}{3}} & \frac{1}{\sqrt{3}} \\ \frac{1}{\sqrt{2}} & \frac{1}{2}\sqrt{\frac{2}{3}} & -\frac{1}{\sqrt{3}} \end{pmatrix} \begin{pmatrix} \frac{2}{\sqrt{2}} & \frac{1}{\sqrt{2}} & \frac{1}{\sqrt{2}} \\ 0 & \sqrt{\frac{3}{2}} & \frac{1}{2}\sqrt{\frac{2}{3}} \\ 0 & 0 & \frac{2}{\sqrt{3}} \end{pmatrix}.$$

(ii)
$$\begin{pmatrix} \frac{1}{2} & \frac{1}{2\sqrt{51}} \\ \frac{1}{2} & \frac{9}{2\sqrt{51}} \\ \frac{1}{2} & \frac{1}{2\sqrt{51}} \\ -\frac{1}{2} & \frac{11}{2\sqrt{51}} \end{pmatrix} \begin{pmatrix} 2 & -\frac{1}{2} \\ 0 & \frac{51}{2\sqrt{51}} \end{pmatrix}.$$

(iii)
$$\simeq \begin{pmatrix} 0.500 & 0.070 & -0.530 \\ 0.500 & 0.630 & -0.175 \\ 0.500 & 0.070 & 0.822 \\ -0.500 & 0.770 & 0.117 \end{pmatrix} \begin{pmatrix} 2.000 & -0.500 & 0.000 \\ 0.000 & 3.571 & -0.560 \\ 0.000 & 0.000 & 3.699 \end{pmatrix}.$$

Chapter 19

Singular Value Decomposition

In this chapter, we shall develop another type of factorization, which is a generalization of the diagonalization procedure discussed in Chapter 18. This factorization is applicable to any real $m \times n$ matrix A, and in the literature has been named as the *singular value decomposition (SVD)*. Besides solving linear systems, SVD has a wide variety of applications in diverse fields such as data compression, noise reduction, storage, estimating the rank of a matrix, and transmission of digitized information. Before we state the main result of this chapter, let us recall the following steps:

S1 For an $m \times n$ matrix A, the $m \times m$ matrix AA^t and the $n \times n$ matrix $A^t A$ are symmetric.

S2 In view of Problem 17.8, the eigenvalues of $A^t A$ are real and nonnegative. We assume that the eigenvalues $\lambda_1, \cdots, \lambda_r$ are nonzero and arrange them in decreasing order, i.e., $\lambda_1 \geq \lambda_2 \geq \cdots \lambda_r > 0$.

S3 In view of Theorem 18.2 the matrix $A^t A$ has n orthonormal eigenvectors $v^1, \cdots, v^n$. Let $v^1, \cdots, v^r$ be those corresponding to the eigenvalues $\lambda_1, \cdots, \lambda_r$ of $A^t A$. Clearly, an immediate extension of Problem 11.3 implies that r is the rank of $A^t A$, which is the same as that of A.

S4 Since $AA^t Av^i = A(\lambda_i v^i) = \lambda_i Av^i$, it follows that v^i is an eigenvector of $A^t A$, which implies that Av^i is an eigenvector of AA^t, with the same eigenvalue λ_i.

S5 From Problem 17.8 it follows that $\|Av^i\|_2 = \sigma_i \|v^i\|_2$, where $\sigma_i = \lambda_i^2$, $i = 1, \cdots, r$. These σ_i are called the *singular values* of A. Clearly, $\sigma_1 \geq \sigma_2 \geq \cdots \geq \sigma_r > 0$.

S6. Define $u^i = Av^i/\sigma_i$, $i = 1, \cdots, r$. Then, clearly

$$(u^i, u^i) = \frac{1}{\lambda_i}(Av^i, Av^i) = \frac{\lambda_i}{\lambda_i}(v^i, v^i) = 1,$$

i.e., the vectors u^i, $i = 1, \cdots, r$ are also normalized. Further,we have

$$(u^i)^t Av^j = (Av^i/\sigma_i)^t Av^j = \frac{1}{\sigma_i}(v^i)^t A^t Av^j = \frac{1}{\sigma_i}(v^i)^t \lambda_j v^j = \frac{\lambda_j}{\sigma_i}(v^i)^t v^j,$$

which implies that for all $1 \leq i, j \leq r$, $(u^i)^t Av^j = 0$ for $i \neq j$, and σ_i for $i = j$. This means that the set $\{u^1, \cdots, u^r\}$ is an orthonormal basis for the column space A.

S7.   The matrix $V = (v^1, \cdots, v^r | v^{r+1}, \cdots, v^n)$ orthogonally diagonalizes $A^t A$; we use Remark 14.2 to extend the set $\{u^1, \cdots, u^r\}$ to $\{u^1, \cdots, u^r | u^{r+1}, \cdots, u^m\}$, which forms an orthonormal basis for R^m, and construct the matrix $U = (u^1, \cdots, u^r | u^{r+1}, \cdots, u^m)$; we define the matrix Σ as follows

$$\Sigma = \left(\begin{array}{ccc} D_{rr} & \vdots & 0_{r,n-r} \\ \cdots\cdots & \vdots & \cdots\cdots \\ 0_{m-r,r} & \vdots & 0_{m-r,n-r} \end{array} \right);$$

here, D_{rr} is the diagonal matrix

$$D_{rr} = \left(\begin{array}{cccc} \sigma_1 & 0 & \cdots & 0 \\ 0 & \sigma_2 & \cdots & 0 \\ \vdots & \vdots & \ddots & \cdots \\ 0 & 0 & \cdots & \sigma_r \end{array} \right)$$

and $0_{k\ell}$ is the $k \times \ell$ zero matrix.

S8.   We claim that $U\Sigma = AV$. Indeed, we have

$$\begin{aligned} U\Sigma &= (\sigma_1 u^1, \cdots, \sigma_r u^r | 0, \cdots, 0) \\ &= (Av^1, \cdots, Av^r | Av^{r+1}, \cdots, Av^n) = AV. \end{aligned}$$

S9.   Using the orthogonality of V, it follows that

$$A = U\Sigma V^t. \tag{19.1}$$

Theorem 19.1.   Every $m \times n$ matrix A with rank r has a singular value decomposition, i.e, (19.1) holds.

Proof.   The steps S1–S9 provide the constructive proof of (19.1).   ∎

It is clear that for a symmetric matrix, singular value decomposition is the same as orthogonal diagonalization, provided the eigenvalues are arranged in a decreasing order.

Example 19.1.   We shall find singular value decomposition of the matrix

$$A = \left(\begin{array}{cccc} 2 & 2 & 1 & 0 \\ 1 & -1 & 0 & 1 \end{array} \right).$$

For this, we note that

$$A^t A = \left(\begin{array}{cccc} 5 & 3 & 2 & 1 \\ 3 & 5 & 2 & -1 \\ 2 & 2 & 1 & 0 \\ 1 & -1 & 0 & 1 \end{array} \right).$$

For the matrix $A^t A$, eigenvalues and the corresponding eigenvectors are $\lambda_1 = 9, \lambda_2 = 3, \lambda_3 = 0, \lambda_4 = 0$ and $(2, 2, 1, 0)^t, (1 - 1, 0, 1)^t, (-1 - 1, 4, 0)^t, (-1, 1, 0, 2)^t$. Thus, $\sigma_1 = 3, \sigma_2 = \sqrt{3}$,

$$\Sigma = \begin{pmatrix} 3 & 0 & 0 & 0 \\ 0 & \sqrt{3} & 0 & 0 \end{pmatrix} \quad \text{and} \quad V = \begin{pmatrix} \frac{2}{3} & \frac{1}{\sqrt{3}} & -\frac{1}{3\sqrt{2}} & -\frac{1}{\sqrt{6}} \\ \frac{2}{3} & -\frac{1}{\sqrt{3}} & -\frac{1}{3\sqrt{2}} & \frac{1}{\sqrt{6}} \\ \frac{1}{3} & 0 & \frac{4}{3\sqrt{2}} & 0 \\ 0 & \frac{1}{\sqrt{3}} & 0 & \frac{2}{\sqrt{6}} \end{pmatrix}.$$

We also compute

$$u^1 = Av^1/\sigma_1 = \begin{pmatrix} 1 \\ 0 \end{pmatrix} \quad \text{and} \quad u^2 = Av^2/\sigma_2 = \begin{pmatrix} 0 \\ 1 \end{pmatrix},$$

and hence

$$U = \begin{pmatrix} 1 & 0 \\ 0 & 1 \end{pmatrix}.$$

From (19.1), now it follows that

$$\begin{pmatrix} 2 & 2 & 1 & 0 \\ 1 & -1 & 0 & 1 \end{pmatrix} = \begin{pmatrix} 1 & 0 \\ 0 & 1 \end{pmatrix} \begin{pmatrix} 3 & 0 & 0 & 0 \\ 0 & \sqrt{3} & 0 & 0 \end{pmatrix} \begin{pmatrix} \frac{2}{3} & \frac{2}{3} & \frac{1}{3} & 0 \\ \frac{1}{\sqrt{3}} & -\frac{1}{\sqrt{3}} & 0 & \frac{1}{\sqrt{3}} \\ -\frac{1}{3\sqrt{2}} & -\frac{1}{3\sqrt{2}} & \frac{4}{3\sqrt{2}} & 0 \\ -\frac{1}{\sqrt{6}} & \frac{1}{\sqrt{6}} & 0 & \frac{2}{\sqrt{6}} \end{pmatrix}.$$

Remark 19.1. From (19.1) it immediately follows that $A^t = V \Sigma^t U^t$. Thus, from Example 19.1, we have the following factorization:

$$\begin{pmatrix} 2 & 1 \\ 2 & -1 \\ 1 & 0 \\ 0 & 1 \end{pmatrix} = \begin{pmatrix} \frac{2}{3} & \frac{1}{\sqrt{3}} & -\frac{1}{3\sqrt{2}} & -\frac{1}{\sqrt{6}} \\ \frac{2}{3} & -\frac{1}{\sqrt{3}} & -\frac{1}{3\sqrt{2}} & \frac{1}{\sqrt{6}} \\ \frac{1}{3} & 0 & \frac{4}{3\sqrt{2}} & 0 \\ 0 & \frac{1}{\sqrt{3}} & 0 & \frac{2}{\sqrt{6}} \end{pmatrix} \begin{pmatrix} 3 & 0 \\ 0 & \sqrt{3} \\ 0 & 0 \\ 0 & 0 \end{pmatrix} \begin{pmatrix} 1 & 0 \\ 0 & 1 \end{pmatrix}.$$

The same factorization can be directly obtained by following the above steps S1–S9.

Singular value decomposition connects four fundamental spaces of A in a natural way. We state and prove this result in the following theorem.

Theorem 19.2. Let the $m \times n$ matrix A have the rank r, and let $U \Sigma V^t$ be its singular value decomposition; then the following hold:

(i)   the set $\{u^1, \cdots, u^r\}$ is an orthonormal basis for $C(A)$

(ii) the set $\{u^{r+1}, \cdots, u^m\}$ is an orthonormal basis for $C(A)^\perp = \mathcal{N}(A^t)$

(iii)   the set $\{v^1, \cdots, v^r\}$ is an orthonormal basis for $R(A)$

(iv)   the set $\{v^{r+1}, \cdots, v^n\}$ is an orthonormal basis for $R(A)^\perp = \mathcal{N}(A)$.

Proof.   (i)   It has already been shown in S6. (ii)   Since from S7, $\{u^1, \cdots, u^r | u^{r+1}, \cdots, u^m\}$ extends $\{u^1, \cdots, u^r\}$ to an orthonormal basis of R^m, it follows that each vector in $\{u^{r+1}, \cdots, u^m\}$ is orthogonal to the $\text{Span}\{u^1, \cdots, u^r\} = C(A)$. Thus, $\{u^{r+1}, \cdots, u^m\}$ is an orthonormal set of $m - r$ vectors in $C(A)^\perp = \mathcal{N}(A^t)$. Now from Corollary 11.4, we have $n(A^t) = m - r$, which implies that $\{u^{r+1}, \cdots, u^m\}$ is an orthonormal basis for $\mathcal{N}(A^t)$. (iii) and (iv)   The set of eigenvectors $\{v^1, \cdots, v^r | v^{r+1}, \cdots, v^n\}$ of $A^t A$ is an orthonormal set. These are ordered so that the first r eigenvectors correspond to the positive eigenvalues and the remaining $n - r$ to the zero eigenvalue. Thus, $\{v^{r+1}, \cdots, v^n\}$ is an orthonormal set of $n - r$ vectors in the null space of $A^t A$, which in view of Problem 11.2 is the same as the null space of A. Now from Theorem 11.4 we know that $n(A)$ is $n - r$, thus the set $\{v^{r+1}, \cdots, v^n\}$ must be an orthonormal basis for $\mathcal{N}(A)$. Next, since each vector of the set $\{v^{r+1}, \cdots, v^n\}$ is orthogonal to each vector of the set $\{v^1, \cdots, v^r\}$, it follows that each vector of the set is orthogonal to the $\text{Span}\{v^{r+1}, \cdots, v^n\} = \mathcal{N}A)$. But this shows that $\{v^1, \cdots, v^r\}$ is an orthonormal set of r vectors in $\mathcal{N}(A)^\perp = R(A)$. Finally, since $R(A)$ has dimension r, the set $\{v^1, \cdots, v^r\}$ must be an orthonormal basis for $R(A)$.   ∎

Remark 19.2.   Following the patricians shown in S7 for the matrices $U, \Sigma,$ and V, we can block multiply $U\Sigma V^t$ to get the *reduced singular value decomposition* of the matrix A as

$$A = U_1 D_{rr} V_1^t = (u^1, \cdots, u^r) \begin{pmatrix} \sigma_1 & 0 & \cdots & 0 \\ 0 & \sigma_2 & \cdots & 0 \\ \vdots & \vdots & \ddots & \cdots \\ 0 & 0 & \cdots & \sigma_r \end{pmatrix} \begin{pmatrix} (v^1)^t \\ \vdots \\ (v^r)^t \end{pmatrix}. \quad (19.2)$$

Remark 19.3.   Multiplying the right side of (19.2), we obtain the reduced singular value expansion of the matrix A as

$$A = \sigma_1 u^1 (v^1)^t + \cdots + \sigma_r u^r (v^r)^t. \quad (19.3)$$

Example 19.2.   In view of Example 19.1 and Remarks 19.2 and 19.3, we have the reduced singular value decomposition

$$\begin{pmatrix} 2 & 2 & 1 & 0 \\ 1 & -1 & 0 & 1 \end{pmatrix} = \begin{pmatrix} 1 & 0 \\ 0 & 1 \end{pmatrix} \begin{pmatrix} 3 & 0 \\ 0 & \sqrt{3} \end{pmatrix} \begin{pmatrix} \frac{2}{3} & \frac{2}{3} & \frac{1}{3} & 0 \\ \frac{1}{\sqrt{3}} & -\frac{1}{\sqrt{3}} & 0 & \frac{1}{\sqrt{3}} \end{pmatrix}$$

and the reduced singular value expansion

$$\begin{pmatrix} 2 & 2 & 1 & 0 \\ 1 & -1 & 0 & 1 \end{pmatrix} = 3 \begin{pmatrix} 1 \\ 0 \end{pmatrix} \left(\frac{2}{3}, \frac{2}{3}, \frac{1}{3}, 0 \right) + \sqrt{3} \begin{pmatrix} 0 \\ 1 \end{pmatrix} \left(\frac{1}{\sqrt{3}}, -\frac{1}{\sqrt{3}}, 0, \frac{1}{\sqrt{3}} \right).$$

Example 19.3. In view of Remarks 19.1–19.3, we have the reduced singular value decomposition

$$\begin{pmatrix} 2 & 1 \\ 2 & -1 \\ 1 & 0 \\ 0 & 1 \end{pmatrix} = \begin{pmatrix} \frac{2}{3} & \frac{1}{\sqrt{3}} \\ \frac{2}{3} & -\frac{1}{\sqrt{3}} \\ \frac{1}{3} & 0 \\ 0 & \frac{1}{\sqrt{3}} \end{pmatrix} \begin{pmatrix} 3 & 0 \\ 0 & \sqrt{3} \end{pmatrix} \begin{pmatrix} 1 & 0 \\ 0 & 1 \end{pmatrix}$$

and the reduced singular value expansion

$$\begin{pmatrix} 2 & 1 \\ 2 & -1 \\ 1 & 0 \\ 0 & 1 \end{pmatrix} = 3 \begin{pmatrix} \frac{2}{3} \\ \frac{2}{3} \\ \frac{1}{3} \\ 0 \end{pmatrix} (1,0) + \sqrt{3} \begin{pmatrix} \frac{1}{\sqrt{3}} \\ -\frac{1}{\sqrt{3}} \\ 0 \\ \frac{1}{\sqrt{3}} \end{pmatrix} (0,1).$$

Remark 19.4. When the matrix A is invertible, i.e., $r = n$, then $\sigma_i > 0$, $i = 1, \cdots, n$ and the singular value decomposition of A takes the form

$$A = (u^1, \cdots, u^n) \begin{pmatrix} \sigma_1 & 0 & \cdots & 0 \\ 0 & \sigma_2 & \cdots & 0 \\ \vdots & \vdots & \ddots & \cdots \\ 0 & 0 & \cdots & \sigma_n \end{pmatrix} \begin{pmatrix} (v^1)^t \\ \vdots \\ (v^n)^t \end{pmatrix}. \tag{19.4}$$

Example 19.4. In Example 18.3, we have seen that the matrix A in Example 16.4 cannot be diagonalized; however, it has a singular value decomposition given by

$$\begin{pmatrix} 2 & 1 & -1 \\ -3 & -1 & 1 \\ 9 & 3 & -4 \end{pmatrix} \simeq \begin{pmatrix} -0.2192 & -0.7939 & -0.5671 \\ 0.2982 & -0.6079 & 0.7359 \\ -0.9290 & -0.0078 & 0.3700 \end{pmatrix}$$
$$\times \begin{pmatrix} 11.0821 & 0 & 0 \\ 0 & 0.3442 & 0 \\ 0 & 0 & 0.2621 \end{pmatrix} \begin{pmatrix} -0.8747 & -0.2982 & 0.3820 \\ 0.4826 & -0.6079 & 0.6305 \\ -0.0442 & -0.7359 & -0.6757 \end{pmatrix}.$$

Problems

19.1. Use steps S1–S9 to show that

$$\begin{pmatrix} 1 & 2 & 1 & 0 \\ 2 & 0 & 1 & 1 \end{pmatrix} = \begin{pmatrix} \frac{1}{\sqrt{2}} & \frac{-1}{\sqrt{2}} \\ \frac{1}{\sqrt{2}} & \frac{1}{\sqrt{2}} \end{pmatrix} \begin{pmatrix} 3 & 0 & 0 & 0 \\ 0 & \sqrt{3} & 0 & 0 \end{pmatrix} \begin{pmatrix} \frac{1}{\sqrt{2}} & \frac{2}{3\sqrt{2}} & \frac{2}{3\sqrt{2}} & \frac{1}{3\sqrt{2}} \\ \frac{1}{\sqrt{6}} & \frac{-2}{\sqrt{6}} & 0 & \frac{1}{\sqrt{6}} \\ 0 & \frac{1}{3} & \frac{-2}{3} & \frac{2}{3} \\ \frac{-1}{\sqrt{3}} & 0 & \frac{1}{\sqrt{3}} & \frac{1}{\sqrt{3}} \end{pmatrix}$$

and

$$
\begin{pmatrix} 1 & 2 \\ 2 & 0 \\ 1 & 1 \\ 0 & 1 \end{pmatrix} = \begin{pmatrix} \frac{1}{\sqrt{2}} & \frac{1}{\sqrt{6}} & 0 & \frac{-1}{\sqrt{3}} \\ \frac{2}{3\sqrt{2}} & \frac{-2}{\sqrt{6}} & \frac{1}{3} & 0 \\ \frac{2}{3\sqrt{2}} & 0 & \frac{-2}{3} & \frac{1}{\sqrt{3}} \\ \frac{1}{3\sqrt{2}} & \frac{1}{\sqrt{6}} & \frac{2}{3} & \frac{1}{\sqrt{3}} \end{pmatrix} \begin{pmatrix} 3 & 0 \\ 0 & \sqrt{3} \\ 0 & 0 \\ 0 & 0 \end{pmatrix} \begin{pmatrix} \frac{1}{\sqrt{2}} & \frac{1}{\sqrt{2}} \\ \frac{-1}{\sqrt{2}} & \frac{1}{\sqrt{2}} \end{pmatrix}.
$$

19.2. Use steps S1–S9 to show that

$$
\begin{pmatrix} 1 & 3 \\ -3 & 3 \\ -3 & 1 \\ 1 & 1 \end{pmatrix} = \begin{pmatrix} \frac{\sqrt{2}}{2\sqrt{7}} & \frac{\sqrt{2}}{\sqrt{3}} \\ \frac{3\sqrt{2}}{2\sqrt{7}} & 0 \\ \frac{\sqrt{2}}{\sqrt{7}} & -\frac{\sqrt{2}}{2\sqrt{3}} \\ 0 & \frac{\sqrt{2}}{2\sqrt{3}} \end{pmatrix} \begin{pmatrix} 2\sqrt{7} & 0 \\ 0 & 2\sqrt{3} \end{pmatrix} \begin{pmatrix} -\frac{1}{\sqrt{2}} & \frac{1}{\sqrt{2}} \\ \frac{1}{\sqrt{2}} & \frac{1}{\sqrt{2}} \end{pmatrix}
$$

and

$$
\begin{pmatrix} 1 & -3 & -3 & 1 \\ 3 & 3 & 1 & 1 \end{pmatrix} = \begin{pmatrix} -\frac{1}{\sqrt{2}} & \frac{1}{\sqrt{2}} \\ \frac{1}{\sqrt{2}} & \frac{1}{\sqrt{2}} \end{pmatrix} \begin{pmatrix} 2\sqrt{7} & 0 \\ 0 & 2\sqrt{3} \end{pmatrix}
$$
$$
\times \begin{pmatrix} \frac{\sqrt{2}}{2\sqrt{7}} & \frac{3\sqrt{2}}{2\sqrt{7}} & \frac{\sqrt{2}}{\sqrt{7}} & 0 \\ \frac{\sqrt{2}}{\sqrt{3}} & 0 & -\frac{\sqrt{2}}{2\sqrt{3}} & \frac{\sqrt{2}}{2\sqrt{3}} \end{pmatrix}.
$$

19.3. Use steps S1–S9 to show that

$$
\begin{pmatrix} -1 & 0 & 1 \\ -1 & 1 & 2 \\ 0 & 1 & 1 \end{pmatrix} = \begin{pmatrix} \frac{1}{6}\sqrt{6} & \frac{-1}{2}\sqrt{2} \\ \frac{1}{3}\sqrt{6} & 0 \\ \frac{1}{6}\sqrt{6} & \frac{1}{2}\sqrt{2} \end{pmatrix} \begin{pmatrix} 3 & 0 & 0 \\ 0 & 1 & 0 \end{pmatrix} \begin{pmatrix} \frac{-1}{6}\sqrt{6} & \frac{1}{6}\sqrt{6} & \frac{1}{3}\sqrt{6} \\ \frac{1}{2}\sqrt{2} & \frac{1}{2}\sqrt{2} & 0 \\ \frac{1}{3}\sqrt{3} & \frac{-1}{3}\sqrt{3} & \frac{1}{3}\sqrt{3} \end{pmatrix}.
$$

19.4. Use steps S1–S9 to show that

$$
\begin{pmatrix} 1 & 2 & 3 \\ 2 & 1 & 0 \\ 1 & 1 & 2 \\ 0 & 3 & 4 \end{pmatrix} \simeq \begin{pmatrix} -0.559 & -0.132 & -0.298 \\ -0.140 & -0.895 & 0.413 \\ -0.353 & -0.211 & -0.710 \\ -0.738 & 0.370 & 0.487 \end{pmatrix}
$$
$$
\times \begin{pmatrix} 6.667 & 0.000 & 0.000 \\ 0.000 & 2.249 & 0.000 \\ 0.000 & 0.000 & 0.700 \end{pmatrix} \begin{pmatrix} -0.179 & -0.573 & -0.800 \\ -0.949 & -0.116 & 0.295 \\ -0.261 & 0.811 & -0.523 \end{pmatrix}.
$$

Chapter 20

Differential and Difference Systems

In this chapter we shall show how linear algebra (especially eigenvalues and eigenvectors) plays an important role to find the solutions of homogeneous differential and difference systems with constant coefficients. Such systems occur in a wide variety of real world applications.

We recall that a linear homogeneous differential system with constant coefficients appears as

$$u' = Au, \qquad (20.1)$$

where $A = (a_{ij})$ is an $n \times n$ given matrix with constant elements, and $u = u(x) = (u_1(x), \cdots, u_n(x))^t$ is the column vector of unknown functions. A solution $\phi(x)$ of (20.1) is a column vector valued function $u = \phi(x) = (\phi_1(x), \cdots, \phi_n(x))^t$ of differentiable functions that satisfies (20.1), i.e., $\phi'(x) = A\phi(x)$. Clearly, $u = 0$ is always a solution of (20.1). This solution is known as the *trivial solution* or *zero solution*. Often, we are interested in finding the solution of (20.1) that satisfies the initial condition

$$u(x_0) = u^0. \qquad (20.2)$$

Differential system (20.1) together with the initial condition (20.2) is called an *initial value problem*, and it always has a unique solution.

Example 20.1. For the system

$$u' = Au = \begin{pmatrix} 2 & 1 & 0 \\ 1 & 3 & 1 \\ 0 & 1 & 2 \end{pmatrix} u \qquad (20.3)$$

each of the following column vectors is a solution:

$$\phi^1(x) = \begin{pmatrix} 1 \\ -1 \\ 1 \end{pmatrix} e^x, \quad \phi^2(x) = \begin{pmatrix} -1 \\ 0 \\ 1 \end{pmatrix} e^{2x}, \quad \phi^3(x) = \begin{pmatrix} 1 \\ 2 \\ 1 \end{pmatrix} e^{4x}. \quad (20.4)$$

Also, for (20.3) with the initial condition $u(0) = (1, 2, 3)^t$, the unique solution $\phi(x)$ is

$$\phi(x) = \frac{2}{3} \begin{pmatrix} 1 \\ -1 \\ 1 \end{pmatrix} e^x + \begin{pmatrix} -1 \\ 0 \\ 1 \end{pmatrix} e^{2x} + \frac{4}{3} \begin{pmatrix} 1 \\ 2 \\ 1 \end{pmatrix} e^{4x}. \qquad (20.5)$$

171

The following results are basic for the system (20.1):

T1. If $\phi^1(x), \cdots, \phi^m(x)$ are solutions of (20.1) and $c_1, \cdots, c_m$ are arbitrary constants, then the linear combination $\phi(x) = c_1\phi^1(x) + \cdots + c_m\phi^m(x)$ is also a solution of (20.1).

T2. There exist n linearly independent solutions (see Chapter 8) of (20.1).

T3. The set $S = \{\phi^1(x), \cdots, \phi^n(x)\}$ of n linearly independent solutions is called a *fundamental set of solutions* of (20.1), the matrix of these solutions $\Phi(x) = (\phi^1(x), \cdots, \phi^n(x))$ is called the *fundamental matrix solution* of (20.1), and it satisfies the matrix differential system $\Phi'(x) = A\Phi(x)$, and the linear combination $\phi(x) = c_1\phi^1(x) + \cdots + c_n\phi^n(x) = \Phi(x)c$, where $c = (c_1, \cdots, c_n)^t$ is called the general solution of (20.1). Any solution of (20.1) can be expressed as a unique linear combination of the solutions in the set S.

For the system (20.3), solutions given in (20.4) are linearly independent (see Chapter 8, $W(\phi^1, \phi^2, \phi^3)(0) \neq 0$). Thus, the general solution of (20.3) appears as

$$\phi(x) = c_1 \begin{pmatrix} 1 \\ -1 \\ 1 \end{pmatrix} e^x + c_2 \begin{pmatrix} -1 \\ 0 \\ 1 \end{pmatrix} e^{2x} + c_3 \begin{pmatrix} 1 \\ 2 \\ 1 \end{pmatrix} e^{4x}$$

$$= \begin{pmatrix} e^x & -e^{2x} & e^{4x} \\ -e^x & 0 & 2e^{4x} \\ e^x & e^{2x} & e^{4x} \end{pmatrix} \begin{pmatrix} c_1 \\ c_2 \\ c_3 \end{pmatrix},$$

and from this the solution (20.5) of the initial value problem (20.3), $u(0) = (1, 2, 3)^t$, can be obtained immediately.

Theorem 20.1. Let the matrix A be diagonalizable (see Theorem 18.1). Then, the set

$$\phi^1(x) = u^1 e^{\lambda_1 x}, \cdots, \phi^n(x) = u^n e^{\lambda_n x} \tag{20.6}$$

is a fundamental set of solutions of (20.1). Here, $\lambda_1, \cdots, \lambda_n$ are the eigenvalues (not necessarily distinct) of A and $u^1, \cdots, u^n$ are the corresponding linearly independent eigenvectors.

Proof. Since u^i is an eigenvector of A corresponding to the eigenvalue λ_i, we find

$$(\phi^i(x))' = (u^i e^{\lambda_i x})' = \lambda_i u^i e^{\lambda_i x} = A u^i e^{\lambda_i x} = A\phi^i(x)$$

and hence $\phi^i(x)$ is a solution of (20.1). To show that (20.6) is a fundamental set, we note that $W(0) = \det(u^1, \cdots, u^n) \neq 0$, since $u^1, \cdots, u^n$ are linearly independent. ∎

Example 20.2. From Example 16.1 and Theorem 20.1 it is clear that the column vectors $\phi^1(x), \phi^2(x), \phi^3(x)$ given in (20.4) are the solutions of the system (20.3). Similarly, from Example 16.5 and Theorem 20.1 it follows that for the system

$$u' = Au = \begin{pmatrix} -1 & 0 & 4 \\ 0 & -1 & 2 \\ 0 & 0 & 1 \end{pmatrix} u$$

three linearly independent solutions are

$$\phi^1(x) = \begin{pmatrix} 1 \\ 0 \\ 0 \end{pmatrix} e^{-x}, \quad \phi^2(x) = \begin{pmatrix} 0 \\ 1 \\ 0 \end{pmatrix} e^{-x}, \quad \phi^3(x) = \begin{pmatrix} 2 \\ 1 \\ 1 \end{pmatrix} e^x.$$

Remark 20.1. The general solution of (20.1) can be written as

$$\phi(x) = \sum_{i=1}^{n} c_i e^{\lambda_i x} u^i = (u^1, \cdots, u^n) D(x)(c_1, \cdots, c_n)^t,$$

where $D(x)$ is the diagonal matrix

$$D(x) = \begin{pmatrix} e^{\lambda_1 x} & 0 & \cdots & 0 \\ 0 & e^{\lambda_2 x} & \cdots & 0 \\ \vdots & \vdots & \ddots & \vdots \\ 0 & 0 & \cdots & e^{\lambda_n x} \end{pmatrix}.$$

Since the matrix $P = (u^1, \cdots, u^n)$ is nonsingular, P^{-1} exists, and thus, we can choose the vector $c = (c_1, \cdots, c_n)^t$ as $P^{-1}w$, where w is an arbitrary column vector. Hence, when A is diagonalizable, the general solution of (20.1) in matrix form appears as

$$\phi(x) = PD(x)P^{-1}w. \tag{20.7}$$

Now for an arbitrary $n \times n$ matrix A, we introduce the $n \times n$ matrix e^A as follows:

$$e^A = \sum_{k=0}^{\infty} \frac{A^k}{k!} = I + A + \frac{1}{2!}A^2 + \frac{1}{3!}A^3 + \cdots. \tag{20.8}$$

This exponential series converges for any matrix A. Indeed, from the definition of convergence in Chapter 17, we have

$$\left\| \sum_{k=0}^{m+p} \frac{A^k}{k!} - \sum_{k=0}^{m} \frac{A^k}{k!} \right\| = \left\| \sum_{k=m+1}^{m+p} \frac{A^k}{k!} \right\| \leq \sum_{k=m+1}^{m+p} \frac{\|A^k\|}{k!} \leq \sum_{k=m+1}^{m+p} \frac{\|A\|^k}{k!} \leq e^{\|A\|}.$$

Hence, for any $n \times n$ matrix A, e^A is a well defined $n \times n$ matrix.

Let λ be an eigenvalue of the diagonalizable matrix A and u be the corresponding eigenvector, then from (20.8) and P11 (in Chapter 16), we have

$$e^A u = \sum_{k=0}^{\infty} \left(\frac{A^k}{k!} \right) u = \sum_{k=0}^{\infty} \left(\frac{\lambda^k}{k!} \right) u = e^{\lambda} u,$$

which shows that e^λ is an eigenvalue of the matrix e^A, and u is the corresponding eigenvector. Thus, from Problem 16.1(ii) and (iii), we find

$$\det e^A = \prod_{i=1}^{n} e^{\lambda_i} = e^{\lambda_1 + \cdots + \lambda_n} = e^{\mathrm{Tr}\, A} \neq 0$$

(this identity in fact holds for an any $n \times n$ matrix), and hence the matrix e^A is nonsingular, and therefore diagonalizable.

Summarizing the above considerations, we find that if A is diagonalizable, then a diagonalization of e^{Ax} is $PD(x)P^{-1}$, and hence (20.7) can be written as

$$\phi(x) = e^{Ax}w. \tag{20.9}$$

We also note that, since $u^0 = u(x_0) = e^{Ax_0}w$ gives $w = e^{-Ax_0}u^0$, the solution of (20.1), (20.2) can be written as

$$\phi(x) = e^{Ax}e^{-Ax_0}u^0 = e^{A(x-x_0)}u^0. \tag{20.10}$$

Now we claim that for any $n \times n$ matrix A, $\Phi(x) = e^{Ax}$ is a fundamental matrix solution of (20.1). For this, it suffices to note that

$$\Phi'(x) = Ae^{Ax} = A\Phi(x) \quad \text{and} \quad \Phi(0) = e^{A0} = I.$$

Thus, (20.9) is not restricted to diagonalizable matrices only, but rather provides the general solution of (20.1) for any $n \times n$ matrix A. However, when the matrix A is not diagonalizable, the computation of e^{Ax} is not straightforward. Among the several known methods to compute e^{Ax}, the following seems to be the easiest and most popular.

Theorem 20.2 (Putzer's algorithm). Let $\lambda_1, \cdots, \lambda_n$ be the eigenvalues of the matrix A, which are arranged in some arbitrary but specified order. Then,

$$e^{Ax} = \sum_{j=0}^{n-1} r_{j+1}(x)P_j, \tag{20.11}$$

where $P_0 = I$, $P_j = \prod_{k=1}^{j}(A - \lambda_k I)$, $j = 1, \cdots, n$ and $r_1(x), \cdots, r_n(x)$ are recursively given by

$$r_1'(x) = \lambda_1 r_1(x), \quad r_1(0) = 1$$
$$r_j'(x) = \lambda_j r_j(x) + r_{j-1}(x), \quad r_j(0) = 0, \quad j = 2, \cdots, n.$$

(Note that each eigenvalue in the list is repeated according to its multiplicity. Further, since the matrices $(A - \lambda_i I)$ and $(A - \lambda_j I)$ commute, we can for convenience adopt the convention that $(A - \lambda_j I)$ follows $(A - \lambda_i I)$ if $i > j$.)

Proof. It suffices to show that $\Phi(x)$ defined by $\Phi(x) = \sum_{j=0}^{n-1} r_{j+1}(x)P_j$

satisfies $\Phi'(x) = A\Phi(x)$, $\Phi(0) = I$. For this, we define $r_0(x) \equiv 0$. Then, it follows that

$$
\begin{aligned}
\Phi'(x) - \lambda_n \Phi(x) &= \sum_{j=0}^{n-1}(\lambda_{j+1}r_{j+1}(x) + r_j(x))P_j - \lambda_n \sum_{j=0}^{n-1} r_{j+1}(x)P_j \\
&= \sum_{j=0}^{n-1}(\lambda_{j+1} - \lambda_n)r_{j+1}(x)P_j + \sum_{j=0}^{n-1} r_j(x)P_j \\
&= \sum_{j=0}^{n-2}(\lambda_{j+1} - \lambda_n)r_{j+1}(x)P_j + \sum_{j=0}^{n-2} r_{j+1}(x)P_{j+1} \\
&= \sum_{j=0}^{n-2}\{(\lambda_{j+1}-\lambda_n)P_j + (A - \lambda_{j+1}I)P_j\} r_{j+1}(x) \quad (20.12) \\
&= (A - \lambda_n I)\sum_{j=0}^{n-2} P_j r_{j+1}(x) \\
&= (A - \lambda_n I)(\Phi(x) - r_n(x)P_{n-1}) \\
&= (A - \lambda_n I)\Phi(x) - r_n(x)P_n, \quad (20.13)
\end{aligned}
$$

where to obtain (20.12) and (20.13) we have used $P_{j+1} = (A - \lambda_{j+1}I)P_j$ and $P_n = (A - \lambda_n I)P_{n-1}$, respectively. Now by the Cayley–Hamilton theorem (Chapter 16), $P_n = p(A) = 0$, and therefore (20.13) reduces to $\Phi'(x) = A\Phi(x)$. Finally, to complete the proof we note that $\Phi(0) = \sum_{j=0}^{n-1} r_{j+1}(0)P_j = r_1(0)I = I$. ∎

Example 20.3. Consider a 3×3 matrix A having all the three eigenvalues equal to λ_1. To use Theorem 20.2, we note that $r_1(x) = e^{\lambda_1 x}$, $r_2(x) = xe^{\lambda_1 x}$, $r_3(x) = (1/2)x^2 e^{\lambda_1 x}$ is the solution set of the system

$$
\begin{aligned}
r_1' &= \lambda_1 r_1, & r_1(0) &= 1 \\
r_2' &= \lambda_1 r_2 + r_1, & r_2(0) &= 0 \\
r_3' &= \lambda_1 r_3 + r_2, & r_3(0) &= 0.
\end{aligned}
$$

Thus, it follows that

$$
e^{Ax} = e^{\lambda_1 x}\left[I + x(A - \lambda_1 I) + \frac{1}{2}x^2(A - \lambda_1 I)^2\right]. \quad (20.14)
$$

In particular, the matrix

$$
A = \begin{pmatrix} 2 & 1 & -1 \\ -3 & -1 & 1 \\ 9 & 3 & -4 \end{pmatrix}
$$

has all its eigenvalues equal to -1, and hence from (20.14) we obtain

$$e^{Ax} = \frac{1}{2}e^{-x} \begin{pmatrix} 2+6x-3x^2 & 2x & -2x+x^2 \\ -6x & 2 & 2x \\ 18x-9x^2 & 6x & 2-6x+3x^2 \end{pmatrix}.$$

Example 20.4. Consider a 3×3 matrix A with eigenvalues $\lambda_1, \lambda_1, \lambda_2$. To use Theorem 20.2, we note that $r_1(x) = e^{\lambda_1 x}$, $r_2(x) = xe^{\lambda_1 x}$,

$$r_3(x) = \frac{xe^{\lambda_1 x}}{(\lambda_1-\lambda_2)} + \frac{e^{\lambda_2 x}-e^{\lambda_1 x}}{(\lambda_1-\lambda_2)^2}$$

and hence

$$e^{Ax} = e^{\lambda_1 x}\left[I + x(A-\lambda_1 I) + \left\{\frac{x}{(\lambda_1-\lambda_2)} + \frac{e^{(\lambda_2-\lambda_1)x}-1}{(\lambda_1-\lambda_2)^2}\right\}(A-\lambda_1 I)^2\right]. \tag{20.15}$$

In particular, the matrix

$$A = \begin{pmatrix} -1 & 0 & 4 \\ 0 & -1 & 2 \\ 0 & 0 & 1 \end{pmatrix}$$

has the eigenvalues $-1, -1, 1$ and hence from (20.15) we find

$$e^{Ax} = \begin{pmatrix} e^{-x} & 0 & 2(e^x-e^{-x}) \\ 0 & e^{-x} & e^x-e^{-x} \\ 0 & 0 & e^x \end{pmatrix}.$$

Next, we shall consider the difference system

$$u(k+1) = Au(k), \quad k \in \mathbb{N} = \{0,1,2,\cdots\} \tag{20.16}$$

where again $A = (a_{ij})$ is an $n \times n$ given matrix with constant elements, and $u = u(k) = (u_1(k), \cdots, u_n(k))^t$ is the column vector of unknown functions defined for all nonnegative integers. A solution $\phi(k)$, $k \in \mathbb{N}$ of (20.16) is a column vector valued function $\phi(k) = (\phi_1(k), \cdots, \phi_n(k))^t$ that satisfies (20.16), i.e., $\phi(k+1) = A\phi(k)$. Clearly, $u = 0$ is always a solution of (20.16). This solution is known as a *trivial solution* or the *zero solution*. Often, we are interested in finding the solution of (20.16) that satisfies the initial condition

$$u(0) = u^0. \tag{20.17}$$

Difference system (20.16) together with the initial condition (20.17) is called an *initial value problem*, and it always has a unique solution. It is clear that any (general) solution of (20.16) can be written as

$$\phi(k) = A^k c, \quad k \in \mathbb{N} \tag{20.18}$$

where c is an arbitrary column vector. From (20.18) it immediately follows that the unique solution of (20.16), (20.17) is $\phi(k) = A^k u^0$, $k \in \mathbb{N}$. Now when A is diagonalizable we can state the following result whose proof is similar to that of Theorem 20.1.

Theorem 20.3. Let the notations and hypotheses of Theorem 20.1 be satisfied. Then, the set

$$\phi^1(k) = u^1 \lambda_1^k, \cdots, \phi^n(k) = u^n \lambda_n^k, \quad k \in \mathbb{N}$$

is a fundamental set of solutions of (20.16).

Further, we note that like (20.1) the general solution of (20.16) can be written as

$$\phi(k) = A^k c = \sum_{i=1}^{n} c_i \lambda_i^k u^i = PD(k)P^{-1}c, \qquad (20.19)$$

where again $P = (u^1, \cdots, u^n)$ is the nonsingular matrix, $c = (c_1, \cdots, c_n)^t$ is the constant vector, and $D(k)$ is the diagonal matrix

$$D(k) = \begin{pmatrix} \lambda_1^k & 0 & \cdots & 0 \\ 0 & \lambda_2^k & \cdots & 0 \\ \vdots & \vdots & \ddots & \vdots \\ 0 & 0 & \cdots & \lambda_n^k \end{pmatrix}.$$

From (20.19) it follows that $A^k = PD(k)P^{-1}$ provided the matrix A is diagonalizable (see Chapter 18).

Example 20.5. For the discrete system (20.16) with the same matrix as in (20.3), the fundamental set of solutions is

$$\phi^1(k) = \begin{pmatrix} 1 \\ -1 \\ 1 \end{pmatrix}, \quad \phi^2(k) = \begin{pmatrix} -1 \\ 0 \\ 1 \end{pmatrix} 2^k, \quad \phi^3(k) = \begin{pmatrix} 1 \\ 2 \\ 1 \end{pmatrix} 4^k.$$

Example 20.6. For the discrete system (20.16) with the same matrix as in Example 20.2, the fundamental set of solutions is

$$\phi^1(k) = \begin{pmatrix} 1 \\ 0 \\ 0 \end{pmatrix} (-1)^k, \quad \phi^2(x) = \begin{pmatrix} 0 \\ 1 \\ 0 \end{pmatrix} (-1)^k, \quad \phi^3(x) = \begin{pmatrix} 2 \\ 1 \\ 1 \end{pmatrix}.$$

If the matrix A is not diagonalizable we compute A^k by using the discrete version of Theorem 20.2.

Theorem 20.4 (discrete Putzer's algorithm). Let the notations and hypotheses of Theorem 20.2 be satisfied. Then, for all $k \in \mathbb{N}$

$$A^k = \sum_{j=0}^{n-1} w_{j+1}(k) P_j, \qquad (20.20)$$

where

$$
\begin{aligned}
w_1(k+1) &= \lambda_1 w_1(k), \quad w_1(0) = 1 \\
w_j(k+1) &= \lambda_j w_j(k) + w_{j-1}(k), \quad w_j(0) = 0, \quad j = 2, \cdots, n.
\end{aligned}
\tag{20.21}
$$

Proof. Differentiating (20.11) k times and substituting $x = 0$ gives (20.20), where $w_j(k) = r_j^{(k)}(0)$, $1 \leq j \leq n$. ∎

Remark 20.2. The solution of the system (20.21) is

$$
\begin{aligned}
w_1(k) &= \lambda_1^k \\
w_j(k) &= \sum_{\ell=0}^{k-1} \lambda_j^{k-1-\ell} w_{j-1}(\ell), \quad j = 2, \cdots, n.
\end{aligned}
$$

Example 20.7. For a 3×3 matrix A having all the three eigenvalues equal to λ_1 it follows from (20.14) that

$$
A^k = \lambda_1^k I + k\lambda_1^{k-1}(A - \lambda_1 I) + \frac{1}{2}k(k-1)\lambda_1^{k-2}(A - \lambda_1 I)^2.
\tag{20.22}
$$

In particular, for the matrix A in Example 20.3, we have

$$
A^k = \frac{1}{2}(-1)^k \begin{pmatrix} 2 - 3k - 3k^2 & -2k & k + k^2 \\ 6k & 2 & -2k \\ -9k - 9k^2 & -6k & 2 + 3k + 3k^2 \end{pmatrix}.
$$

Example 20.8. For a 3×3 matrix A with eigenvalues $\lambda_1, \lambda_1, \lambda_2$ it follows from (20.15) that

$$
A^k = \lambda_1^k I + k\lambda_1^{k-1}(A - \lambda_1 I) + \left\{ \frac{k\lambda_1^{k-1}}{(\lambda_1 - \lambda_2)} + \frac{\lambda_2^k - \lambda_1^k}{(\lambda_1 - \lambda_2)^2} \right\}(A - \lambda_1 I)^2.
\tag{20.23}
$$

In particular, for the matrix A in Example 20.4, we have

$$
A^k = \begin{pmatrix} (-1)^k & 0 & 2(1 - (-1)^k) \\ 0 & (-1)^k & (1 - (-1)^k) \\ 0 & 0 & 1 \end{pmatrix}.
$$

Problems

20.1. (i) If $A = \begin{pmatrix} \alpha & \beta \\ -\beta & \alpha \end{pmatrix}$, show that

$$
e^{Ax} = e^{\alpha x} \begin{pmatrix} \cos \beta x & \sin \beta x \\ -\sin \beta x & \cos \beta x \end{pmatrix}.
$$

(ii)   If $A = \begin{pmatrix} 0 & 1 \\ -1 & -2\delta \end{pmatrix}$, show that

$$e^{Ax} = \begin{pmatrix} e^{-\delta x}\left(\cos\omega x + \dfrac{\delta}{\omega}\sin\omega x\right) & \dfrac{1}{\omega}e^{-\delta x}\sin\omega x \\[3mm] -\dfrac{1}{\omega}e^{-\delta x}\sin\omega x & e^{-\delta x}\left(\cos\omega x - \dfrac{\delta}{\omega}\sin\omega x\right) \end{pmatrix},$$

where $\omega = \sqrt{1-\delta^2}$.

(iii) If

$$A = \begin{pmatrix} 0 & 1 & 0 & 0 \\ 3\omega^2 & 0 & 0 & 2\omega \\ 0 & 0 & 0 & 1 \\ 0 & -2\omega & 0 & 0 \end{pmatrix},$$

show that

$$e^{Ax} = \begin{pmatrix} 4 - 3\cos\omega x & \dfrac{1}{\omega}\sin\omega x & 0 & \dfrac{2}{\omega}(1-\cos\omega x) \\[3mm] 3\omega\sin\omega x & \cos\omega x & 0 & 2\sin\omega x \\[3mm] 6(-\omega x + \sin\omega x) & -\dfrac{2}{\omega}(1-\cos\omega x) & 1 & \dfrac{1}{\omega}(-3\omega x + 4\sin\omega x) \\[3mm] 6\omega(-1+\cos\omega x) & -2\sin\omega x & 0 & -3 + 4\cos\omega x \end{pmatrix}.$$

(iv) If $A^2 = \alpha A$, show that $e^{Ax} = I + [(e^{\alpha x}-1)/\alpha]A$.

20.2. Let A and P be $n \times n$ matrices given by

$$A = \begin{pmatrix} \lambda & 1 & 0 & \cdots & 0 \\ 0 & \lambda & 1 & \cdots & 0 \\ & \cdots & & & \\ 0 & 0 & 0 & \cdots & 1 \\ 0 & 0 & 0 & \cdots & \lambda \end{pmatrix}, \quad P = \begin{pmatrix} 0 & 1 & 0 & 0 & \cdots & 0 \\ 0 & 0 & 1 & 0 & \cdots & 0 \\ & \cdots & & & & \\ 0 & 0 & 0 & 0 & \cdots & 1 \\ 0 & 0 & 0 & 0 & \cdots & 0 \end{pmatrix}.$$

Show that

(i) $P^n = 0$

(ii) $(\lambda I)P = P(\lambda I)$

(iii) $e^{Ax} = e^{\lambda x}\left[I + xP + \dfrac{1}{2!}x^2 P^2 + \cdots + \dfrac{1}{(n-1)!}x^{n-1}P^{n-1}\right].$

20.3. Let A and B be two $n \times n$ similar matrices, i.e., (13.16) holds. Show that

(i) $u(x)$ is a solution of (20.1) if and only if $v(x) = Pu(x)$ is a solution of the differential system $v' = Bv$.

(ii) $e^{Ax} = P^{-1}e^{Bx}P$.

20.4. Find the general solution of the differential system (20.1), where the matrix A is given by

(i) $\begin{pmatrix} 4 & -2 \\ 5 & 2 \end{pmatrix}$ (ii) $\begin{pmatrix} 7 & 6 \\ 2 & 6 \end{pmatrix}$ (iii) $\begin{pmatrix} 0 & 1 & 1 \\ 1 & 0 & 1 \\ 1 & 1 & 0 \end{pmatrix}$ (iv) $\begin{pmatrix} 1 & -1 & 4 \\ 3 & 2 & -1 \\ 2 & 1 & -1 \end{pmatrix}$

(v) $\begin{pmatrix} -1 & 1 & 0 \\ 0 & -1 & 0 \\ 0 & 0 & 3 \end{pmatrix}$ (vi) $\begin{pmatrix} 5 & -3 & -2 \\ 8 & -5 & -4 \\ -4 & 3 & 3 \end{pmatrix}$.

20.5. Find the general solution of the difference system (20.16) for each matrix A given in Problem 20.4.

Answers or Hints

20.1. Verify directly.
20.2. (i) Observe that in each multiplication the position of 1 is shifted by one column, so in P^2 the nth and $(n-1)$th rows are 0.
(ii) Obvious.
(iii) Since $A = \lambda I + P$, we can use Parts (i) and (ii).
20.3. (i) Verify directly.
(ii) $e^{Ax} = e^{P^{-1}BPx}$ now expand the right side.
20.4. (i) $e^{3x} \begin{pmatrix} 2\cos 3x & 2\sin 3x \\ \cos 3x + 3\sin 3x & \sin 3x - 3\cos 3x \end{pmatrix} \begin{pmatrix} c_1 \\ c_2 \end{pmatrix}$.

(ii) $\begin{pmatrix} 2e^{10x} & 3e^{3x} \\ e^{10x} & -2e^{3x} \end{pmatrix} \begin{pmatrix} c_1 \\ c_2 \end{pmatrix}$.

(iii) $\begin{pmatrix} e^{2x} & e^{-x} & 0 \\ e^{2x} & 0 & e^{-x} \\ e^{2x} & -e^{-x} & -e^{-x} \end{pmatrix} \begin{pmatrix} c_1 \\ c_2 \\ c_3 \end{pmatrix}$.

(iv) $\begin{pmatrix} -e^x & e^{-2x} & e^{3x} \\ 4e^x & -e^{-2x} & 2e^{3x} \\ e^x & -e^{-2x} & e^{3x} \end{pmatrix} \begin{pmatrix} c_1 \\ c_2 \\ c_3 \end{pmatrix}$.

(v) $\begin{pmatrix} 0 & -e^{-x} & xe^{-x} \\ 0 & 0 & e^{-x} \\ e^{3x} & 0 & 0 \end{pmatrix} \begin{pmatrix} c_1 \\ c_2 \\ c_3 \end{pmatrix}$.

(vi) $e^x \begin{pmatrix} 1 & 0 & 2x \\ 0 & 2 & 4x \\ 2 & -3 & -2x-1 \end{pmatrix} \begin{pmatrix} c_1 \\ c_2 \\ c_3 \end{pmatrix}$.

20.5. (i) $(3\sqrt{2})^k \begin{pmatrix} \cos\frac{k\pi}{4} + \frac{1}{3}\sin\frac{k\pi}{4} & -\frac{2}{3}\sin\frac{k\pi}{4} \\ \frac{5}{3}\sin\frac{k\pi}{4} & \cos\frac{k\pi}{4} - \frac{1}{3}\sin\frac{k\pi}{4} \end{pmatrix} \begin{pmatrix} c_1 \\ c_2 \end{pmatrix}$.

(ii) $\frac{1}{7}\begin{pmatrix} 4(10)^k + 3(3)^k & 6(10)^k - 6(3)^k \\ 2(10)^k - 2(3)^k & 3(10)^k + 4(3)^k \end{pmatrix}\begin{pmatrix} c_1 \\ c_2 \end{pmatrix}.$

(iii) $\frac{1}{3}\begin{pmatrix} 2^k + 2(-1)^k & 2^k - (-1)^k & 2^k - (-1)^k \\ 2^k - (-1)^k & 2^k + 2(-1)^k & 2^k - (-1)^k \\ 2^k - (-1)^k & 2^k - (-1)^k & 2^k + 2(-1)^k \end{pmatrix}\begin{pmatrix} c_1 \\ c_2 \\ c_3 \end{pmatrix}.$

(iv) $\frac{1}{6}\begin{pmatrix} -1 & (-2)^k & 3^k \\ 4 & -(-2)^k & 2(3)^k \\ 1 & -(-2)^k & 3^k \end{pmatrix}\begin{pmatrix} -1 & 2 & -3 \\ 2 & 2 & -6 \\ 3 & 0 & 3 \end{pmatrix}\begin{pmatrix} c_1 \\ c_2 \\ c_3 \end{pmatrix}.$

(v) $\begin{pmatrix} (-1)^k & k(-1)^{k-1} & 0 \\ 0 & (-1)^k & 0 \\ 0 & 0 & 3^k \end{pmatrix}\begin{pmatrix} c_1 \\ c_2 \\ c_3 \end{pmatrix}.$

(vi) $\begin{pmatrix} 1+4k & -3k & -2k \\ 8k & 1-6k & -4k \\ -4k & 3k & 1+2k \end{pmatrix}\begin{pmatrix} c_1 \\ c_2 \\ c_3 \end{pmatrix}.$

Chapter 21

Least Squares Approximation

We know that the $m \times n$ system (5.2) has a solution if and only if $r(A) = r(A : b)$, i.e., $b \in C(A)$. However, in a wide range of applications we encounter problems in which b may not be in $C(A)$. For such a problem we seek a vector(s) $\hat{x} \in R^n$ so that the *error* $\|A\hat{x} - b\|_2$ is as small as possible (minimized), i.e.,

$$\|A\hat{x} - b\|_2 \leq \|Ax - b\|_2 \qquad (21.1)$$

for all $x \in R^n$. This solution(s) $\hat{x}$ is called *the least squares approximate solution*. We emphasize that to find approximate solutions to such problems several different error criteria and the corresponding numerical procedures have been proposed, but among all these the method of least squares approximations is the simplest to implement. We shall provide two different proofs to the following important theorem.

Theorem 21.1. The set of least squares solutions to the system (5.2) is given by the consistent $n \times n$ system (known as *normal equations*)

$$A^t Ax = A^t b. \qquad (21.2)$$

If the columns of A are linearly independent, then there exists a unique least squares solution to (5.2) and it is given by

$$\hat{x} = (A^t A)^{-1} A^t b. \qquad (21.3)$$

If the columns of A are linearly dependent, then there are an infinite number of least squares solutions.

First Proof. Consider the *error function* $E(x) = Ax - b$, $x \in R^n$. Clearly, we need to find a vector(s) x that minimizes $\|E(x)\|_2$. For this, we consider the scalar function

$$
\begin{aligned}
\Phi(x) = \|E(x)\|_2^2 &= (Ax - b, Ax - b) \\
&= (Ax - b)^t (Ax - b) \\
&= x^t A^t Ax - x^t A^t b - b^t Ax + b^t b \\
&= x^t A^t Ax - 2x^t A^t b + b^t b.
\end{aligned} \qquad (21.4)
$$

Now for $0 \neq h \in R^n$, we have

$$
\begin{aligned}
\Phi(x+h) - \Phi(x) &= x^t A^t A h + h^t A^t A x + h^t A^t A h - 2h^t A^t b \\
&= h^t A^t A h + 2h^t A^t A x - 2h^t A^t b \qquad (21.5) \\
&= \|Ah\|_2^2 + 2h^t (A^t A x - A^t b).
\end{aligned}
$$

Clearly, $\Phi(x)$ attains a relative minimum (maximum) provided $\Phi(x+h)-\Phi(x)$ remains of fixed sign for small values of the vector h, i.e., $\|h\|_2$ is small. However, in (21.5) the sign of $\Phi(x+h) - \Phi(x)$ depends on the term $h^t(A^t A x - A^t b)$, and hence for an extremum it is necessary that $A^t A x - A^t b = 0$. Further, if this condition is satisfied for $x = \hat{x}$, then (21.5) reduces to

$$
\Phi(\hat{x}+h) - \Phi(\hat{x}) = \|Ah\|_2^2 \geq 0,
$$

which ensures that $\Phi(x)$ indeed has a minimum at $\hat{x}$. In conclusion, the least-squares solutions of (5.2) are the solutions of the system (21.2).

Next, we shall show that the system (21.2) is consistent. For this, we note that

$$
A^t b = \begin{pmatrix} a_{11} \\ \vdots \\ a_{1n} \end{pmatrix} b_1 + \cdots + \begin{pmatrix} a_{m1} \\ \vdots \\ a_{mn} \end{pmatrix} b_m
$$

is a linear combination of the columns of A^t, and hence $A^t b \in C(A^t)$. Since $C(A^t) = C(A^t A)$ (see Problem 11.3), we find that $A^t b \in C(A^t A)$, which in turn implies that $r(A^t A) = r(A^t A \mid A^t b)$, i.e., the system (21.2) is consistent. Now from Problem 11.3 it follows that

$$
r(A) = r(A^t A) = r(A^t A \mid A^t b). \qquad (21.6)
$$

Finally, from Problem 11.2, we have $\mathcal{N}(A) = \mathcal{N}(A^t A)$, and hence from Theorem 11.4 and (21.6) it follows that $\dim \mathcal{N}(A) = n - r(A) = n - r(A^t A) = \dim \mathcal{N}(A^t A)$. Thus, if $r(A) = n$, then $\mathcal{N}(A^t A) = 0$, which implies that $A^t A$ is nonsingular, and $(A^t A)^{-1}$ exists. Therefore, in this case a unique solution of the least squares problem is given by (21.3).

Second Proof. Let $S = C(A)$. In view of Theorem 14.5 and Remark 14.4, the vector b can be written as $b = b^1 + b^2$, where $b_1 \in S$ is the orthogonal projection of b onto S and $b^2 \in S^\perp$ is the component of b orthogonal to S. Now as in the First Proof it follows that

$$
\begin{aligned}
\|Ax - b\|_2 &= \|(Ax - b^1) - b^2\|_2 \\
&= \|Ax - b^1\|_2 - 2(Ax - b^1, b^2) + \|b^2\|_2.
\end{aligned}
$$

Since Ax and b^1 are in S and b^2 is in $S^\perp$, the middle term vanishes, and thus we have

$$
\|Ax - b\|_2 = \|Ax - b^1\|_2 + \|b^2\|_2.
$$

Clearly, the right side is minimized if x is a solution of the system

$$Ax = b^1. \tag{21.7}$$

Since b^1 is in S, this system is consistent. Any solution of this system denoted as $\hat{x}$ is a least squares solution of (5.2). Further, this solution is unique provided the columns of A are lineraly independent.

Now suppose that $\hat{x}$ is a solution of the system (21.7). Since b^2 is orthogonal to the columns of A, it follows that $A^t b^2 = 0$. Thus, we have

$$A^t A\hat{x} = A^t b^1 = A^t(b - b^2) = A^t b,$$

i.e., $\hat{x}$ is a solution of the system (21.2). Conversely, suppose that $\hat{x}$ is a solution of (21.2). Then, we have

$$A^t(b - A\hat{x}) = 0.$$

This means that the vector $(b - A\hat{x})$ is orthogonal to each row of A^t, i.e., to each column of A. Since $S = C(A)$, we conclude that $(b - A\hat{x}) \in S^\perp$. Hence, the vector b can be written as

$$b = A\hat{x} + (b - A\hat{x}),$$

where $A\hat{x} \in S$ and $(b - A\hat{x}) \in S^\perp$. But again in view of Theorem 14.5 and Remark 14.4 such a decomposition is unique, and hence $A\hat{x} = b^1$. ∎

Corollary 21.1. Let A be an $m \times n$ matrix with $r(A) = n$, and suppose $A = QR$ is a QR factorization of A (see Theorem 18.3). Then, the upper triangular system

$$Rx = Q^t b \tag{21.8}$$

gives the least squares approximate solution of the system (5.2).

Proof. In view of Theorem 18.3, the system (21.2) can be written as $R^t Q^t Q R x = R^t Q^t b$. However, since $Q^t Q = I$ and R is invertible, this system is the same as (21.8). ∎

Example 21.1. For the system

$$\begin{pmatrix} 1 & 1 & 2 \\ 1 & 2 & 3 \\ 1 & 3 & 4 \\ 3 & 4 & 9 \end{pmatrix} \begin{pmatrix} x_1 \\ x_2 \\ x_3 \end{pmatrix} = \begin{pmatrix} 2 \\ 2 \\ h \\ k \end{pmatrix} \tag{21.9}$$

the echelon form of the augmented matrix appears as

$$\left(\begin{array}{ccc|c} 1 & 1 & 2 & 2 \\ 0 & 1 & 1 & 0 \\ 0 & 0 & 2 & k - 6 \\ 0 & 0 & 0 & h - 2 \end{array} \right).$$

Thus, the system has a solution if and only if $h = 2$, and in such a case the solution is

$$\left(\frac{1}{2}(10 - k), \frac{1}{2}(6 - k), -\frac{1}{2}(6 - k)\right)^t. \tag{21.10}$$

Since the columns of the matrix A in (21.9) are linearly independent, from Corollary 21.1 and Example 18.6, it follows that the unique least squares solution of (21.9) can be obtained by solving the system

$$\begin{pmatrix} \frac{6}{\sqrt{3}} & \frac{9}{\sqrt{3}} & \frac{18}{\sqrt{3}} \\ 0 & \frac{3}{\sqrt{3}} & \frac{2}{\sqrt{3}} \\ 0 & 0 & \frac{2}{\sqrt{6}} \end{pmatrix} \begin{pmatrix} \hat{x}_1 \\ \hat{x}_2 \\ \hat{x}_3 \end{pmatrix} = \begin{pmatrix} \frac{1}{2\sqrt{3}} & -\frac{1}{2\sqrt{3}} & -\frac{2}{\sqrt{6}} \\ \frac{1}{2\sqrt{3}} & \frac{1}{2\sqrt{3}} & -\frac{1}{\sqrt{6}} \\ \frac{1}{2\sqrt{3}} & \frac{3}{2\sqrt{3}} & 0 \\ \frac{3}{2\sqrt{3}} & -\frac{1}{2\sqrt{3}} & \frac{1}{\sqrt{6}} \end{pmatrix} \begin{pmatrix} 2 \\ 2 \\ h \\ k \end{pmatrix}$$

$$= \begin{pmatrix} \frac{1}{2\sqrt{3}}(4 + h + 3k) \\ \frac{1}{2\sqrt{3}}(3h - k) \\ \frac{1}{\sqrt{6}}(k - 6) \end{pmatrix}.$$

Using backward process this system gives the unique least squares approximate solution of (21.9) as

$$x_3 = -\frac{1}{2}(6 - k), \quad x_2 = \frac{1}{2}(h - k + 4), \quad x_1 = \frac{1}{6}(38 - 4h - 3k). \tag{21.11}$$

Clearly, (21.11) is the same as (21.10) for $h = 2$, as it should be.

For (21.9) the system (21.2) is

$$\begin{pmatrix} 12 & 18 & 36 \\ 18 & 30 & 56 \\ 36 & 56 & 110 \end{pmatrix} \begin{pmatrix} x_1 \\ x_2 \\ x_3 \end{pmatrix} = \begin{pmatrix} 4 + h + 3k \\ 6 + 3h + 4k \\ 10 + 4h + 9k \end{pmatrix},$$

which as expected gives the same solution as (21.11).

Example 21.2. Consider the system

$$\begin{pmatrix} 1 & 3 & 2 \\ 3 & -5 & -1 \\ 4 & 2 & 3 \\ 2 & 2 & 2 \end{pmatrix} \begin{pmatrix} x_1 \\ x_2 \\ x_3 \end{pmatrix} = \begin{pmatrix} 1 \\ 2 \\ 3 \\ 4 \end{pmatrix}. \tag{21.12}$$

In view of Problem 8.1(i), in this system the columns of the matrix A are linearly dependent. Further, for this system the echelon form of the augmented matrix appears as

$$\begin{pmatrix} 1 & 3 & 2 & | & 1 \\ 0 & 1 & \frac{1}{2} & | & \frac{1}{14} \\ 0 & 0 & 0 & | & -\frac{2}{7} \\ 0 & 0 & 0 & | & \frac{16}{7} \end{pmatrix},$$

and hence it has no solution. For (21.12) the system (21.2) is

$$\begin{pmatrix} 30 & 0 & 15 \\ 0 & 42 & 21 \\ 15 & 21 & 18 \end{pmatrix} \begin{pmatrix} x_1 \\ x_2 \\ x_3 \end{pmatrix} = \begin{pmatrix} 27 \\ 7 \\ 17 \end{pmatrix}. \tag{21.13}$$

The system (21.13) has an infinite number of solutions

$$\left(\frac{9}{10} - \frac{1}{2}c, \ \frac{1}{6} - \frac{1}{2}c, \ c \right)^t. \tag{21.14}$$

Thus, the system (21.12) has an infinite number of least squares solutions given by (21.14).

Example 21.3. From Problem 6.3(i), we know that the system

$$\begin{pmatrix} 2 & 7 & 4 & 3 \\ 8 & 5 & 3 & 9 \\ 1 & 3 & 6 & 4 \end{pmatrix} \begin{pmatrix} x_1 \\ x_2 \\ x_3 \\ x_3 \end{pmatrix} = \begin{pmatrix} 1 \\ 3 \\ 7 \end{pmatrix} \tag{21.15}$$

has an infinite number of solutions. For (21.15) the system (21.2) is

$$\begin{pmatrix} 69 & 57 & 38 & 82 \\ 57 & 83 & 61 & 78 \\ 38 & 61 & 61 & 63 \\ 82 & 78 & 63 & 106 \end{pmatrix} \begin{pmatrix} x_1 \\ x_2 \\ x_3 \\ x_4 \end{pmatrix} = \begin{pmatrix} 33 \\ 43 \\ 55 \\ 58 \end{pmatrix}.$$

This system has an infinite number of solutions:

$$\frac{1}{197}(62, -161, 300, 0)^t + \frac{1}{197}(-203, 41, -118, 198)^t c. \tag{21.16}$$

Thus, the system (21.15) has an infinite number of least squares solutions given by (21.16). These solutions are exactly the same as given for Problem 6.3(i), as they should be.

Problems

21.1. Find the least squares solution of the system considered in Example 6.3.

21.2. Find the least squares solution of the system considered in Problem 6.3(ii).

21.3. Find the least squares solution of the system considered in Problem 6.3(v).

21.4. Show that the following system has no solution

$$\begin{pmatrix} 1 & 2 & 1 & 5 \\ 1 & 2 & -1 & 1 \\ 2 & 4 & -3 & 0 \end{pmatrix} \begin{pmatrix} x_1 \\ x_2 \\ x_3 \\ x_4 \end{pmatrix} = \begin{pmatrix} 1 \\ 1 \\ 3 \end{pmatrix}.$$

Find its least sqaures solution.

21.5. Show that the following system has no solution

$$\begin{pmatrix} 2 & 1 & 3 \\ 4 & 3 & 5 \\ 8 & 2 & 16 \\ -4 & 1 & -11 \\ -2 & 3 & -5 \end{pmatrix} \begin{pmatrix} x_1 \\ x_2 \\ x_3 \end{pmatrix} = \begin{pmatrix} 1 \\ 2 \\ 4 \\ 2 \\ 2 \end{pmatrix}.$$

Find its least sqaures solution.

Answers or Hints

21.1. $\left(\frac{1}{3}, \frac{2}{3}, 1\right)^t$.

21.2. $\left(2, \frac{13}{55}, -\frac{53}{55}, 0\right)^t + \left(-5, \frac{48}{55}, \frac{147}{55}, 1\right)^t c.$

21.3. $(0, 1, 1, 0, 0)^t + \left(-\frac{13}{30}, \frac{9}{10}, \frac{16}{15}, 1, 0\right)^t c + \left(\frac{1}{30}, \frac{7}{10}, \frac{8}{15}, 0, 1\right)^t d.$

21.4. That the Solution does not exist follows from the echelon form of

the augmented matrix $\left(\begin{array}{cccc|c} 1 & 2 & 1 & 5 & 1 \\ 0 & 0 & -2 & -4 & 0 \\ 0 & 0 & 0 & 0 & 1 \end{array}\right)$. Its least squares solution is

$\frac{1}{15}(17, 0, -3, 0)^t + (-2, 1, 0, 0)^t c + (-3, 0, -2, 1)^t d.$

21.5. That the Solution does not exist follows from the echelon form

of the augmented matrix $\left(\begin{array}{ccc|c} 2 & 1 & 3 & 1 \\ 0 & 1 & -1 & 0 \\ 0 & 0 & 2 & 0 \\ 0 & 0 & 0 & 4 \\ 0 & 0 & 0 & 3 \end{array}\right)$. Its least squares solution is

$\frac{1}{956}(491, 680, -173)^t.$

Chapter 22

Quadratic Forms

Quadratic forms occur naturally in physics, economics, engineering (control theory), and analytic geometry (quadratic curves and surfaces). Particularly, recall that the equation (quadratic form) of a central quadratic curve in a plane, after translating the origin of the rectangular coordinate system to the centre of the curve, appears as

$$q_2(x,y) = (x,y) \begin{pmatrix} a & b \\ b & c \end{pmatrix} \begin{pmatrix} x \\ y \end{pmatrix} = ax^2 + 2bxy + cy^2 = d.$$

We also know that by rotating the axes (using a proper transformation) this equation (quadratic form) in a new coordinate system can be reduced to a "canonical" (diagonal) form:

$$q_2(x',y') = (x',y') \begin{pmatrix} a' & 0 \\ 0 & c' \end{pmatrix} \begin{pmatrix} x' \\ y' \end{pmatrix} = a'x'^2 + c'y'^2 = d.$$

In this chapter we shall study quadratic forms in n variables $x_1, \cdots, x_n$, i.e.,

$$q_n(x_1, \cdots, x_n) = \sum_{i=1}^{n} b_i x_i^2 + 2 \sum_{i<j}^{n} c_{ij} x_i x_j, \quad b_i, c_{ij} \in R. \tag{22.1}$$

Clearly, in (22.1) each term is of degree two.

In matrix form, (22.1) can be written as

$$q_n(x) = q_n(x_1, \cdots, x_n) = x^t A x, \tag{22.2}$$

where $A = (a_{ij})$ is an $n \times n$ symmetric matrix with $a_{ii} = b_i$, $a_{ij} = a_{ji} = c_{ij}$. In (22.2), if we let $x = Py$, a linear tranformation of the variables, then it follows that

$$q_n(y) = (Py)^t A(Py) = y^t(P^t AP)y, \tag{22.3}$$

i.e, $P^t AP$ provides the matrix representation of q_n in the new variables. Now since A is symmetric, in view of Theorem 18.2, we can always find an orthonormal matrix Q and the diagonal matrix D consisting of the eigenvalues of A such that $D = Q^{-1}AQ = Q^t AQ$. Thus, with the proper choice of the matrix P, (22.3) in the new variables can be reduced to a diagonal form,

$$q_n(y) = y^t D y. \tag{22.4}$$

189

We summarize our above discussion in the following result.

Theorem 22.1. If A is a symmetric matrix, then there exists an orthonormal matrix Q such that the transformation $y = Q^t x$ changes the quadratic form (22.2) into the diagonal quadratic form (22.4).

Example 22.1. Consider the quadratic form

$$q_3(x_1, x_2, x_3) = 11x_1^2 + 11x_2^2 + 14x_3^2 - 2x_1x_2 - 8x_1x_3 - 8x_2x_3, \qquad (22.5)$$

which in matrix form is the same as $q_3(x) = x^t A x$, where the symmetric matrix A is as in (18.1). From Example 18.4 it is clear that (22.5) can be reduced to a diagonal form,

$$q_3(y_1, y_2, y_3) = (y_1, y_2, y_3) \begin{pmatrix} 6 & 0 & 0 \\ 0 & 12 & 0 \\ 0 & 0 & 18 \end{pmatrix} \begin{pmatrix} y_1 \\ y_2 \\ y_3 \end{pmatrix} = 6y_1^2 + 12y_2^2 + 18y_3^2.$$

Here, in view of Remark 18.3, the new varible vector y is

$$y = Q^t x = \begin{pmatrix} \frac{1}{\sqrt{3}} & \frac{1}{\sqrt{3}} & \frac{1}{\sqrt{3}} \\ -\frac{1}{\sqrt{2}} & \frac{1}{\sqrt{2}} & 0 \\ -\frac{1}{\sqrt{6}} & -\frac{1}{\sqrt{6}} & \frac{2}{\sqrt{6}} \end{pmatrix} \begin{pmatrix} x_1 \\ x_2 \\ x_3 \end{pmatrix} = \begin{pmatrix} \frac{1}{\sqrt{3}}(x_1 + x_2 + x_3) \\ \frac{1}{\sqrt{2}}(-x_1 + x_2) \\ \frac{1}{\sqrt{6}}(-x_1 - x_2 + 2x_3) \end{pmatrix}.$$

Example 22.2. Consider the quadratic form

$$\begin{aligned} q_4(x_1, x_2, x_3, x_4) &= x_1^2 + \frac{5}{3}x_2^2 - \frac{5}{6}x_3^2 - \frac{5}{6}x_4^2 + 2x_1x_2 + 2x_1x_3 \\ &\quad + 2x_1x_4 - \frac{8}{3}x_2x_3 - \frac{8}{3}x_2x_4 + \frac{7}{3}x_3x_4, \end{aligned} \qquad (22.6)$$

which in matrix form is the same as $q_4(x) = x^t A x$, where the symmetric matrix A is as in Example 18.5. From this example, it is clear that (22.6) can be reduced to a diagonal form

$$q_4(y_1, q_2, q_3, q_4) = 2y_1^2 + 3y_2^2 - 2y_3^2 - 2y_4^2.$$

Here the new variable vector y is

$$y = Q^t x = \begin{pmatrix} \frac{1}{\sqrt{12}}(3x_1 + x_2 + x_3 + x_4) \\ \frac{1}{\sqrt{6}}(-2x_2 + x_3 + x_4) \\ \frac{1}{\sqrt{6}}(-x_1 + x_2 + 2x_3) \\ \frac{1}{\sqrt{12}}(-x_1 + x_2 - x_3 + 3x_4) \end{pmatrix}.$$

Now we classify the quadratic form (22.2) according as its values: The quadratic form $q_n(x)$ is called *positive definite* if $q_n(x) > 0$, $x \in R^n / \{0\}$ and *positive semidefinite* if $q_n(x) \geq 0$, $x \in R^n$. The quadratic form $q_n(x)$ is called *negative definite* if $q_n(x) < 0$, $x \in R^n/\{0\}$ and *negative semidefinite* if $q_n(x) \leq 0$, $x \in R^n$. The quadratic form $q_n(x)$ is called *indefinite* if it is both positive and negative for $x \in R^n$.

Theorem 22.2. If A is a symmetric matrix, then the quadratic form $q_n(x) = x^t A x$ is positive definite if A has only positive eigenvalues, negative definite if A has only negative eigenvalues, and indefinite if A has both positive and negative eigenvalues.

Proof. If $\lambda_1, \cdots, \lambda_n$ are the eigenvalues of A, then from (22.4) it follows that

$$q_n(y) = \lambda_1 y_1^2 + \cdots + \lambda_n y_n^2, \qquad (22.7)$$

where $y = Q^{-1}x = Q^t x$. Thus, if all eigenvalues of A are positive, then from the fact that $y = 0$ implies $x = 0$ (one-to-one correspondence between y and x), we find $q_n(x) > 0$ for all $x \in R^n/\{0\}$, i.e., $q_n(x)$ is positive definite. If $\lambda_k \leq 0$, we can select $y = e^k$, then $q_n(y) = \lambda_k \leq 0$, i.e., $q_n(x)$ is not positive definite. The other cases can be discussed similarly. ∎

Example 22.3. In view of Theorem 22.2 the quadratic form (22.5) is positive definite, whereas (22.6) is indefinite. From Example 18.1, it is clear that the quadratic form

$$q_3(x_1, x_2, x_3) = -11x_1^2 - 11x_2^2 - 14x_3^2 + 2x_1 x_2 + 8x_1 x_3 + 8x_2 x_3 \quad (22.8)$$

is negative definite (the eigenvalues of the corresponding matrix are $-6, -12, -18$).

The rest of the results in this chapter find maximum and minimum of the quadratic form $q_n(x)$ subject to some constraints, and so belong to a broad field known as *constrained optimization*.

Lemma 22.1. For the quadratic form $q_n(x) = \sum_{i=1}^n b_i x_i^2$ subject to the constraint $\|x\|_2 = 1$, the following hold:

(i)   the maximum value is $b_k = \max\{b_1, \cdots, b_n\}$ and attained at $x = e^k$,

(ii) the minimum value is $b_\ell = \min\{b_1, \cdots, b_n\}$ and attained at $x = e^\ell$.

Proof. Clearly, $q_n(x) = \sum_{i=1}^n b_i x_i^2 \leq b_k \sum_{i=1}^n x_i^2 = b_k$. Further, at $x = e^k$, $q_n(e^k) = b_k$. ∎

Theorem 22.3. Let the symmetric matrix A in (22.2) have the eigenvalues $\lambda_1 \leq \cdots \leq \lambda_n$, and $x^1, \cdots, x^n$ be the corresponding eigenvectors. Then, for the quadratic form $q_n(x) = x^t A x$ subject to the constraint $\|x\|_2 = 1$, the following hold:

(i)   the maximum value is λ_n and attained at $x = x^n/\|x^n\|_2$,

(ii) the minimum value is λ_1 and attained at $x = x^1/\|x^1\|_2$.

Proof. From Theorem 22.1 it follows that the transformation $x = Qy$ reduces (22.2) to the diagonal form (22.7); here, Q consists of vectors that are obtained by orthonormalizing the eigenvectors $x^1, \cdots, x^n$. Now since $y = Q^t x$ and the matrix Q is orthonormal, we have $\|y\|_2^2 = \|Q^t x\|_2^2 = (Q^t x, Q^t x) = (Q^t x)^t (Q^t x) = x^t (QQ^t) x = x^t x = (x, x) = \|x\|_2^2$. Thus, $\|y\|_2 = 1$ if and only if $\|x\|_2 = 1$, and hence $x^t A x$ and $y^t D y$ assume the same set of values as x and y range over the set of all unit vectors. Now since in view of Lemma 22.1 the maximum (minimum) value of $q_n(y)$ is attained at e^n (e^1), for $q_n(x)$ the maximum (minimum) value is attained at $x = Qe^n$ (Qe^1) $= x^n / \|x^n\|_2$ ($x^1 / \|x^1\|_2$). ∎

Example 22.4. From Example 22.1 and Theorem 22.3, it follows that

$$6 \leq 11x_1^2 + 11x_2^2 + 14x_3^2 - 2x_1 x_2 - 8x_1 x_3 - 8x_2 x_3 \leq 18$$

provided $x_1^2 + x_2^2 + x_3^2 = 1$. The left equality holds for $x = \left(\frac{1}{\sqrt{3}}, \frac{1}{\sqrt{3}}, \frac{1}{\sqrt{3}} \right)^t$ and the right inequality holds for $\left(-\frac{1}{\sqrt{6}}, -\frac{1}{\sqrt{6}}, \frac{2}{\sqrt{6}} \right)^t$.

Example 22.5. From Example 22.2 and Theorem 22.3, it follows that

$$-2 \leq \left\{ \begin{array}{c} x_1^2 + \dfrac{5}{3} x_2^2 - \dfrac{5}{6} x_3^2 - \dfrac{5}{6} x_4^2 + 2x_1 x_2 + 2x_1 x_3 \\[2mm] +2x_1 x_4 - \dfrac{8}{3} x_2 x_3 - \dfrac{8}{3} x_2 x_4 + \dfrac{7}{3} x_3 x_4 \end{array} \right\} \leq 3$$

provided $x_1^2 + x_2^2 + x_3^2 + x_4^2 = 1$. From Example 18.5 it follows that the left equality holds for $x = \left(-\frac{1}{\sqrt{6}}, \frac{1}{\sqrt{6}}, \frac{2}{\sqrt{6}}, 0 \right)$, $\left(-\frac{1}{\sqrt{12}}, \frac{1}{\sqrt{12}}, -\frac{1}{\sqrt{12}}, \frac{3}{\sqrt{12}} \right)$, and $\left(0, 0, -\frac{1}{\sqrt{2}}, \frac{1}{\sqrt{2}} \right)$. The right-hand equality holds at $\left(0, -\frac{2}{\sqrt{6}}, \frac{1}{\sqrt{6}}, \frac{1}{\sqrt{6}} \right)$.

Remark 22.1. By letting $x_i = a_i z_i$, $1 \leq i \leq n$ we can transform the optimization problem $q_n(x) = x^t A x$ subject to $\sum_{i=1}^{n} x_i^2 / a_i^2 = 1$ into $q_n(z) = (a_1 z_1, \cdots, a_n z_n) A (a_1 z_1, \cdots, a_n z_n)^t$ subject to $\|z\|_2 = 1$, for which Theorem 22.3 is applicable. In particular, if $a_i = a$, $1 \leq i \leq n$, then it simply reduces to $q_n(z) = a^2 q_n(x)$ subject to $\|x\|_2 = 1$.

Example 22.6. From Remark 22.1 and Example 22.4, it follows that

$$24 \leq 11x_1^2 + 11x_2^2 + 14x_3^2 - 2x_1 x_2 - 8x_1 x_3 - 8x_2 x_3 \leq 72$$

provided $x_1^2 + x_2^2 + x_3^2 = 4$. The left equality holds for $x = \left(\frac{2}{\sqrt{3}}, \frac{2}{\sqrt{3}}, \frac{2}{\sqrt{3}} \right)^t$ and the right inequality holds for $\left(-\frac{2}{\sqrt{6}}, -\frac{2}{\sqrt{6}}, \frac{4}{\sqrt{6}} \right)^t$.

Finally, in this chapter we shall prove the following result.

Theorem 22.4. Let the symmetric matrix A in (22.2) have the eigenvalues $\lambda_1 \leq \cdots \leq \lambda_n$ and associated $u^1, \cdots, u^n$ orthonormal eigenvectors. Then, for the quadratic form $q_n(x) = x^t A x$ subject to the constraints $\|x\|_2 = 1$ and $(x, u^n) = 0$, the following hold:

(i) maximum value is λ_{n-1} and attained at $x = u^{n-1}$,

(ii) minimum value is λ_1 and attained at $x = u^1$.

Proof. From (14.5) for a given $x \in R^n$, we have $x = \sum_{i=1}^n c_i u^i$, where $c_i = (x, u^i)$, $1 \leq i \leq n$, but since $c_n = 0$ it follows that $x = \sum_{i=1}^{n-1} c_i u^i$, which implies that $\|x\|_2^2 = \sum_{i=1}^{n-1} c_i^2 = 1$ (as given). Now we successively have

$$q_n(x) = x^t A x = x^t A \left(\sum_{i=1}^{n-1} c_i u^i \right) = x^t \left(\sum_{i=1}^{n-1} c_i A u^i \right)$$

$$= \left(x, \sum_{i=1}^{n-1} \lambda_i c_i u^i \right) = \left(\sum_{i=1}^{n-1} c_i u^i, \sum_{i=1}^{n-1} \lambda_i c_i u^i \right)$$

$$= \sum_{i=1}^{n-1} \lambda_i c_i^2 \leq \lambda_{n-1} \sum_{i=1}^{n-1} c_i^2 = \lambda_{n-1}.$$

Next, we note that for $x = u^{n-1}$, $c_{n-1} = 1$ and $c_i = 0$, $1 \leq i \leq n - 2$, and hence $q(u^{n-1}) = \lambda_{n-1}$, i.e., $q_n(x)$ attains its maximum at $x = u^{n-1}$. The minimum value of $q_n(x)$ by the constraint $(x, u^n) = 0$ does not change. ∎

Example 22.7. From Example 22.1 and Theorem 22.4, it follows that

$$6 \leq 11x_1^2 + 11x_2^2 + 14x_3^2 - 2x_1x_2 - 8x_1x_3 - 8x_2x_3 \leq 12$$

provided $x_1^2 + x_2^2 + x_3^2 = 1$ and $-x_1 - x_2 + 2x_3 = 0$. The left equality holds for $x = \left(\frac{1}{\sqrt{3}}, \frac{1}{\sqrt{3}}, \frac{1}{\sqrt{3}} \right)^t$ and the right inequality holds for $\left(-\frac{1}{\sqrt{2}}, \frac{1}{\sqrt{2}}, 0 \right)^t$.

The following extension of Theorem 22.4 is immediate.

Theorem 22.5. Let λ_i and u^i, $1 \leq i \leq n$ be as in Theorem 22.4. Then, for the quadratic form $q_n(x) = x^t A x$ subject to the constraints $\|x\|_2 = 1$ and $(x, u^n) = (x, u^{n-1}) = \cdots = (x, u^k) = 0$ the maximum (minimum) value is λ_{k-1} (λ_1) and attained at $x = u^{k-1}$ (u^1).

Problems

22.1. Find the canonical form for the quadratic form $q_3(x) = 2x_1x_2 + x_1x_3 - 2x_2x_3$.

22.2. Find the canonical form for the quadratic form $q_3(x) = x_1^2 + \frac{7}{8}x_2^2 - \frac{7}{8}x_3^2 + 2x_1x_2 + 2x_1x_3 + x_2x_3$.

194   *Chapter 22*

22.3. Show that the quadratic form

(i) $q_3(x) = 5x_1^2 + x_2^2 + 5x_3^2 + 4x_1x_2 - 8x_1x_3 - 4x_2x_3$ is positive definite

(ii) $q_3(x) = 3x_1^2 + x_2^2 + 5x_3^2 + 4x_1x_2 - 8x_1x_3 - 4x_2x_3$ is indefinite

(iii) $q_4(x) = 2x_1x_2 + 2x_1x_3 - 2x_1x_4 - 2x_2x_3 + 2x_2x_4 + 2x_3x_4$ is indefinite.

22.4. Let $q_2(x,y) = ax^2 + 4bxy + cy^2$ be a quadratic form, with $a,b,c \in R$.

(i)   Find the values of a,b,c so that $q_2(x,y)$ is indefinite.

(ii) Find the values of a,b,c so that $q_2(x,y)$ is positive definite.

(iii) Find the values of a,b,c so that $q_2(x,y)$ is negative definite.

22.5. For the quadratic form $q_3(x) = x_1^2 + x_2^2 + 5x_3^2 - 6x_1x_2 + 2x_1x_3 - 2x_2x_3$, find the maximum and minimum subject to the constraint $\|x\|_2 = 1$.

22.6. Consider a rectangle inside the ellipse $9x^2 + 16y^2 = 144$. Find positive values of x and y so that the rectangle has the maximum area.

22.7. For the quadratic form $q_3(x) = 5x_1^2 + 6x_2^2 + 7x_3^2 - 4x_1x_2 + 4x_2x_3$, find the maximum and minimum subject to the constraints $\|x\|_2 = 1$ and $-x_1 + 2x_2 + 2x_3 = 0$.

22.8. For the quadratic form $q_3(x) = \alpha x_1^2 + x_2^2 + x_3^2 + 2x_1x_2 + 2x_1x_3 + 6x_2x_3$, find the values of α so that it is negative definite.

22.9. For the optimization problem (22.2), $\|x\|_2 = 1$ assume that m and M are the minimum and maximum values, respectively. Show that for each number c in the interval $m \leq c \leq M$ there is a unit vector u_c such that $u_c^t A u_c = c$.

Answers or Hints

22.1. We let $x_1 = y_1 + y_2, x_2 = y_1 - y_2, x_3 = y_3$, to obtain $q_3(y) = 2y_1^2 - 2y_2^2 - y_1y_3 + 3y_2y_3$, in the basis $\{f^1 = e^1 + e^2 = (1,1,0)^t, f^2 = e^1 - e^2 = (1,-1,0)^t, f^3 = e^3 = (0,0,1)^t\}$. Clearly, $q_3(y) = (\sqrt{2}y_1 - \frac{1}{2\sqrt{2}}y_3)^2 - \frac{1}{8}y_3^2 - 2y_2^2 + 3y_2y_3$. Denoting $z_1 = \sqrt{2}y_1 - \frac{1}{2\sqrt{2}}y_3, z_2 = y_2, z_3 = y_3$, we find $q_3(z) = z_1^2 - \frac{1}{8}z_2^2 - 2z_3^2 + 3z_2z_3$, in the basis $\{g^1 = \frac{1}{\sqrt{2}}f^1, g^2 = f^2, g^3 = \frac{1}{4}f^1 + f^3\}$. It follows that $q_3(z) = z_1^2 + (\sqrt{2}z_2 - \frac{3}{2\sqrt{2}}z_3)^2 + z_3^2$. Now, we let $w_1 = z_1, w_2 = \sqrt{2}z_2 - \frac{3}{2\sqrt{2}}z_3, w_3 = z_3$, to get $q_3(w) = w_1^2 + w_2^2 + w_3^2$, in the basis $\{h^1 = g^1, h^2 = \frac{1}{\sqrt{2}}g^2, h^3 = \frac{3}{4}g^2 + g^3\}$.
22.2. $q_3(x) = (x_1 + x_2 + x_3)^2 - \frac{1}{8}(x_2 + 4x_3)^2 + \frac{1}{8}x_3^2$.
22.3. (i) The eigenvalues of the corresponding matrix are positive $5 \pm 2\sqrt{6}, 1$.
(ii)   The eigenvalues of the corresponding matrix are positive and negative $9.097835, -0.384043, 0.286208$.

(iii) The eigenvalues of the corresponding matrix are positive and negative $1, 1, 1, -3$.

22.4. The matrix associated to the quadratic form is $A = \begin{pmatrix} a & 2b \\ 2b & c \end{pmatrix}$ with eigenvalues $\lambda_1 = (a + c + \sqrt{(a-c)^2 + 16b^2})/2$ and $\lambda_1 = (a + c - \sqrt{(a-c)^2 + 16b^2})/2$. Since $\lambda_1 \lambda_2 = ac - 4b^2$ and $\lambda_1 + \lambda_2 = a + c$, it follows that:

(i) If $ac - 4b^2 < 0$, then $q_2(x, y)$ is indefinite.
(ii) If $ac - 4b^2 > 0$ and $a + c > 0$, then $q_2(x, y)$ is positive definite.
(iii) If $ac - 4b^2 > 0$ and $a + c < 0$, then $q_2(x, y)$ is negative definite.

22.5. The eigenvalues and eigenvectors of the corresponding matrix are $-2, 3, 6$ and $(1, 1, 0)^t, (-1, 1, 1)^t, (1, -1, 2)^t$. Thus, the maximum is 6 at $\left(\frac{1}{\sqrt{6}}, -\frac{1}{\sqrt{6}}, \frac{2}{\sqrt{6}} \right)^t$ and the minimum is -2 at $\left(\frac{1}{\sqrt{2}}, \frac{1}{\sqrt{2}}, 0 \right)^t$.

22.6. The rectangle's area is $S = 4xy$, thus we need to maximize the quadratic form $q_2(x, y) = 4xy$ subject to the constraint $9x^2 + 16y^2 = 144$. Let $x = 3x_1, y = 4y_1$, then we have the equivalent problem: maximize $q_2(x_1, y_1) = 48x_1 y_1$ subject to the constraint $x_1^2 + y_1^2 = 1$. Clearly, $q_2(x_1, y_1) = (x_1, y_1) \begin{pmatrix} 0 & 24 \\ 24 & 0 \end{pmatrix} (x_1, y_1)^t$, and the matrix A has the eigenvalues $-24, 24$ and the eigenvectors $(-1, 1)^t, (1, 1)^t$. Thus the maximum is 24 and it occurs for $x = 3/\sqrt{2}, y = 4/\sqrt{2}$.

22.7. The eigenvalues and eigenvectors of the corresponding matrix are $3, 6, 9$, and $(-2, -2, 1)^t, (2, -1, 2)^t, (-1, 2, 2)^t$. Since $-x_1 + 2x_2 + 2x_3 = 0$, Theorem 22.4 is applicable. The maximum is 6 at $\left(\frac{2}{3}, -\frac{1}{3}, \frac{2}{3} \right)^t$, and the minimum is 3 at $\left(-\frac{2}{3}, -\frac{2}{3}, \frac{1}{3} \right)^t$.

22.8. For the corresponding matrix the characteristic equation is $(\lambda + 2)[-\lambda^2 + (4 + \alpha)\lambda + (2 - 4\alpha)] = 0$. Thus, $\lambda_1 = -2, \lambda_2 + \lambda_3 = \alpha + 4, \lambda_2 \lambda_3 = 4\alpha - 2$. The quadratic form is negative definite provided $\lambda_2 + \lambda_3 < 0$ ($\alpha < -4$) and $\lambda_2 \lambda_3 > 0$ ($\alpha > 2$). But, then there is no such α.

22.9. Assume that $m < M$. Let u_m and u_M be the unit vectors such that $u_m^t A u_m = m$ and $u_M^t A u_M = M$. Consider the vector
$$u_c = \sqrt{(M - c)/(M - m)} u_m + \sqrt{(c - m)/(M - m)} u_M.$$

Chapter 23

Positive Definite Matrices

Positive definite matrices occur in certain optimization algorithms in mathematical programming, quantum chemistry, and calculation of molecular vibrations. Positive definite matrices are defined only for the symmetric matrices, and in a certain sense are analogues to positive numbers. We begin with the following definition.

Definition 23.1. A symmetric $n \times n$ matrix is called *positive definite* if the quadratic form $q_n(x) = x^t A x$ is positive definite, i.e., $q_n(x) > 0$, $x \in R^n \backslash \{0\}$. Symmetric matrices that are *negative definite* and *indefinite* are defined analogously.

From Theorem 22.2 it follows that a symmetric matrix A is positive definite if it has only positive eigenvalues, negative definite if it has only negative eigenvalues, and indefinite if it has both positive and negative eigenvalues. In the following result we provide another proof of this result.

Theorem 23.1. A symmetric matrix A is positive definite if and only if it has only positive eigenvalues.

Proof. Suppose that A is positive definite. Let λ be an eigenvalue of A and x be the corresponding eigenvector, then we have

$$0 < x^t A x = x^t (\lambda x) = \lambda (x^t x) = \lambda \|x\|_2^2,$$

which immediately implies that $\lambda > 0$. Conversely, assume that all eigenvalues of A are positive. Let $\{x^1, \cdots, x^n\}$ be an orthonormal set of eigenvectors of A, so that any vector $x \in R^n$ can be written as $x = \sum_{i=1}^n c_i x^i$, where $c_i = (x, x^i)$ and $(x, x) = \sum_{i=1}^n c_i^2$. From this, we have

$$
\begin{aligned}
x^t A x &= (x, Ax) = \left(\sum_{i=1}^n c_i x^i, A \sum_{i=1}^n c_i x^i \right) \\
&= \left(\sum_{i=1}^n c_i x^i, \sum_{i=1}^n c_i \lambda_i x^i \right) = \sum_{i=1}^n \lambda_i c_i^2 \geq (\min \lambda_i) \sum_{i=1}^n c_i^2 > 0,
\end{aligned}
$$

and hence A is positive definite. ∎

Example 23.1. In view of Example 18.4, the matrix in (18.1) is positive definite.

Finding eigenvalues of a matrix is not an easy problem. Thus, in what follows we discuss some other criteria.

Theorem 23.2. If A is a symmetric positive definite matrix, then A is nonsingular, in fact, $\det(A) > 0$.

Proof. From Theorem 23.1 and Problem 16.1(iii), we have $\det(A) = \prod_{i=1}^{n} \lambda_i > 0$. We can also show the nonsingularity of A by contradiction: If A is singular, then $Ax = 0$ has a nonzero solution, say, $\tilde{x}$, for which $\tilde{x}^t A \tilde{x} = 0$. But this contradicts the fact that A is positive definite. ∎

Remark 23.1. The converse of Theorem 23.2 does not hold. Indeed, for the matrix $A = \begin{pmatrix} -2 & 0 \\ 0 & -3 \end{pmatrix}$, $\det(A) > 0$, but the eigenvalues are -2 and -3.

Definition 23.1. For an $n \times n$ matrix $A = (a_{ij})$, the *leading principal submatrices* are defined as

$$A^1 = (a_{11}), \quad A^2 = \begin{pmatrix} a_{11} & a_{12} \\ a_{21} & a_{22} \end{pmatrix}, \quad A^3 = \begin{pmatrix} a_{11} & a_{12} & a_{13} \\ a_{21} & a_{22} & a_{23} \\ a_{31} & a_{32} & a_{33} \end{pmatrix}, \cdots, A^n = A.$$

The $\det(A^k)$ is called the k-principal minor of A.

Theorem 23.3. For a symmetric positive definite matrix A, all leading principal submatrices are also positive definite.

Proof. Let $0 \neq x^k \in R^k$, $1 \leq k \leq n$, and set $x = (x^k, 0, \cdots, 0)^t \in R^n$. Then, in view of A being positive definite, we have

$$(x^k)^t A^k x^k = x^t A x > 0.$$

Since $0 \neq x^k \in R^k$ is arbitrary, A^k is positive definite. ∎

In Theorem 7.1 we proved that the matrix A has LU-factorization provided A can be reduced to echelon form without interchanging rows. The following result provides a class of matrices that can be reduced to echelon form without interchanging rows.

Theorem 23.4. If all principal minors $\det(A^k)$, $k = 1, \cdots, n$ of a symmetric matrix A are positive, then A can be reduced to echelon form without interchanging rows. Further, the pivot elements are positive.

Proof. The proof is by induction. If $n = 1$, then $A = (a_{11})$ is in echelon form and $0 < \det(A) = a_{11}$. Assume that the theorem is true for $(n-1) \times (n-1)$

symmetric matrices. We write the $n \times n$ symmetric matrix as

$$
A = \left(
\begin{array}{c|c}
A^{n-1} & \begin{matrix} a_{1n} \\ \vdots \\ a_{(n-1)n} \end{matrix} \\
\hline
a_{n1} \cdots a_{n(n-1)} & a_{nn}
\end{array}
\right).
$$

By inductive hypothesis, we can reduce the matrix A to the form

$$
A^* = \begin{pmatrix}
a_{11}^* & a_{12} & \cdots & a_{1,n-1} & a_{1n} \\
0 & a_{22}^* & \cdots & \tilde{a}_{2,n-1} & \tilde{a}_{2,n} \\
\vdots & \vdots & \ddots & \vdots & \vdots \\
0 & 0 & \cdots & a_{n-1,n-1}^* & \tilde{a}_{n-1,n} \\
a_{n1} & a_{n2} & \cdots & a_{n,n-1} & a_{nn}
\end{pmatrix},
$$

where $a_{ii}^*, i = 1, \cdots, n-1$ are positive. From this it immediately follows that we can reduce A^* and hence A to echelon form without interchanging rows, as

$$
A^{**} = \begin{pmatrix}
a_{11}^* & a_{12} & \cdots & a_{1,n-1} & a_{1n} \\
0 & a_{22}^* & \cdots & \tilde{a}_{2,n-1} & \tilde{a}_{2,n} \\
\vdots & \vdots & \ddots & \vdots & \vdots \\
0 & 0 & \cdots & a_{n-1,n-1}^* & \tilde{a}_{n-1,n} \\
0 & 0 & \cdots & 0 & a_{nn}^*
\end{pmatrix}.
$$

Finally, since $\det(A) > 0$, $\det(A^{n-1}) > 0$, and $\det(A) = \det(A^{**}) = \det(A^{n-1})a_{nn}^*$, it follows that $a_{nn}^* = \det(A)/\det(A^{n-1}) > 0$. ∎

Example 23.2. For the matrix A in (18.1), the echelon form is

$$
\begin{pmatrix}
11 & -1 & -4 \\
0 & \frac{120}{11} & -\frac{48}{11} \\
0 & 0 & \frac{54}{5}
\end{pmatrix}.
$$

For this matrix, it also directly follows that

$$
\begin{aligned}
a_{11}^* &= \det(A^1) = 11 \\
a_{22}^* &= \det(A^2)/\det(A^1) = 120/11 \\
a_{33}^* &= \det(A)/\det(A^2) = 1296/120 = 54/5.
\end{aligned}
$$

Further, this matrix can be LU factorized as

$$
A = LU = \begin{pmatrix}
1 & 0 & 0 \\
-\frac{1}{11} & 1 & 0 \\
-\frac{4}{11} & -\frac{2}{5} & 1
\end{pmatrix}
\begin{pmatrix}
11 & -1 & -4 \\
0 & \frac{120}{11} & -\frac{48}{11} \\
0 & 0 & \frac{54}{5}
\end{pmatrix}.
$$

Combining Remark 7.1 with Theorem 23.4, we have the following result.

Theorem 23.5. If all principal minors $\det(A^k)$, $k = 1, \cdots, n$ of a symmetric matrix A are positive, then A can be uniquely factored as $A = LDL^t$, where L is a lower triangular matrix with all diagonal elements 1, and D is a diagonal matrix with all positive elements.

Proof. In view of Remark 7.1 and Theorem 23.4, the factorization $A = LD\hat{U}$ is unique, where the diagonal matrix D has only positive elements. Now since A is symmetric, we have

$$LD\hat{U} = A = A^t = \hat{U}^t D^t L^t = \hat{U}^t D L^t,$$

which immediately implies that $\hat{U} = L^t$. ∎

Remark 23.2. We denote the diagonal elements of D as $0 < d_{ii}$, $i = 1, \cdots, n$, and define the diagonal matrix $D^{1/2}$ with elements $0 < \sqrt{d_{ii}}$, $i = 1, \cdots, n$. Then, the above factorization can be written as

$$A = LDL^t = (LD^{1/2})(D^{1/2}L^t) = (LD^{1/2})(LD^{1/2})^t = L_c L_c^t,$$

where $L_c = LD^{1/2}$. The factorization $A = L_c L_c^t$ is called *Cholesky decomposition*. Now let us denote by L_{k-1} the $(k-1) \times (k-1)$ upper left corner of L_c, a_k the first $k-1$ entries in column k of A, ℓ_k the first $k-1$ entries in column k of L_c^t, and a_{kk} and ℓ_{kk} the kk entries of A and L_c, respectively. Then, the Cholesky algorithm is:

$$
\begin{aligned}
L_1 &= \sqrt{a_{11}} = \ell_{11} \\
L_{k-1}\ell_k &= a_k, \quad \text{compute} \quad \ell_k \\
\ell_{kk} &= \sqrt{a_{kk} - \ell_k^t \ell_k} \\
L_k &= \begin{pmatrix} L_{k-1} & 0 \\ \ell_k^t & \ell_{kk} \end{pmatrix}, \quad k = 2, \cdots, n.
\end{aligned}
$$

Example 23.3. Using the Cholesky algorithm for the matrix A in (18.1), we successively have

$$\ell_{11} = \sqrt{11}, \quad \ell_{21} = -\frac{1}{\sqrt{11}}, \quad \ell_{22} = \sqrt{\frac{120}{11}},$$

$$\ell_{31} = -\frac{4}{\sqrt{11}}, \quad \ell_{32} = -\sqrt{\frac{96}{55}}, \quad \ell_{33} = \sqrt{\frac{54}{5}}.$$

Thus, the Cholesky decomposition of the matrix A in (18.1) is

$$
\begin{pmatrix} \sqrt{11} & 0 & 0 \\ -\frac{1}{\sqrt{11}} & \sqrt{\frac{120}{11}} & 0 \\ -\frac{4}{\sqrt{11}} & -\sqrt{\frac{96}{55}} & \sqrt{\frac{54}{5}} \end{pmatrix}
\begin{pmatrix} \sqrt{11} & -\frac{1}{\sqrt{11}} & -\frac{4}{\sqrt{11}} \\ 0 & \sqrt{\frac{120}{11}} & -\sqrt{\frac{96}{55}} \\ 0 & 0 & \sqrt{\frac{54}{5}} \end{pmatrix}.
$$

We are now in the position to prove the main result of this chapter.

Theorem 23.6 (Sylvester's criterion). A symmetric matrix A is positive definite if and only if all principal minors $\det(A^k)$, $k = 1, \cdots, n$ are positive.

Proof. If A is positive definite, then all principal minors are positive, as follows from Theorems 23.2 and 23.3. Conversely, the matrix A has the Cholesky decomposition, i.e., $A = L_c L_c^t$. Since $\det(A) > 0$, the matrix L_c^t must be nonsingular, and hence $L_c^t x \neq 0$ for all $x \neq 0$. From this, we find

$$x^t A x = x^t L_c L_c^t x = (L_c^t x)^t (L_c^t x) = \|L_c^t x\|^2 > 0.$$

Hence, the matrix A is positive definite.

Example 23.4. For the matrix A in (18.1), we have $\det(A^1) = 11 > 0$, $\det(A^2) = 120 > 0$, $\det(A^3) = 1296 > 0$; thus in view of Theorem 23.6, the matrix A is positive definite. This in turn implies that the quadratic form in (22.5) is positive definite.

Finally, in this chapter we shall prove the following result.

Theorem 23.7 (polar decomposition). Let the $n \times n$ matrix A have rank r. Then, A can be factored as $A = PQ$, where P is a symmetric $n \times n$ positive semidefinite matrix of rank r, and Q is an $n \times n$ orthogonal matrix. If $r = n$, then the matrix P is positive definite.

Proof. We rewrite the singular value decomposition (19.1) as

$$A = U\Sigma V^t = U\Sigma U^t U V^t = (U\Sigma U^t)(U V^t) = PQ. \tag{23.1}$$

In (23.1), the matrix $Q = UV^t$ is the product of two orthogonal matrices, and hence in view of Problem 4.7, is orthogonal. The matrix $P = U\Sigma U^t$ is symmetric, also orthogonally similar to Σ, and hence in view of P12 (in Chapter 16) P has the same rank and eigenvalues as Σ. This implies that P is positive semidefinite. Clearly, if $r = n$ then all diagonal elements of Σ are positive, and thus P is positive definite. ∎

Example 23.5. For the matrix in Problem 16.2(ii), the singular value decomposition of $U\Sigma V^t$, where

$$U = \begin{pmatrix} -0.0886 & -0.7979 & -0.5963 \\ 0.3080 & -0.5912 & 0.7454 \\ -0.9472 & -0.1176 & 0.2981 \end{pmatrix}, \quad \Sigma = \begin{pmatrix} 23.7448 & 0.0000 & 0.0000 \\ 0.0000 & 2.6801 & 0.0000 \\ 0.0000 & 0.0000 & 0.0000 \end{pmatrix},$$

$$V^t = \begin{pmatrix} -0.7014 & 0.6762 & 0.2254 \\ -0.7128 & -0.6654 & -0.2218 \\ 0.0000 & 0.3162 & -0.9487 \end{pmatrix}$$

and the polar decomposition is PQ, where

$$P = U\Sigma U^t = \begin{pmatrix} 1.8927 & 0.6163 & 2.2442 \\ 0.6163 & 3.1893 & -6.7409 \\ 2.2442 & -6.7409 & 21.3406 \end{pmatrix},$$

$$Q = UV^t = \begin{pmatrix} 0.6309 & 0.2825 & 0.7227 \\ 0.2054 & 0.8373 & -0.5066 \\ 0.7482 & -0.4680 & -0.4702 \end{pmatrix}.$$

Problems

23.1. Show that if a symmetric matrix A is positive definite, then the diagonal elements are positive. Is the converse true?

23.2. Let A be a symmetric positive definite matrix and C be a nonsingular matrix. Show that the matrix $C^t A C$ is also symmetric positive definite.

23.3. Let A and B be $n \times n$ symmetric positive definite matrices. Show that $A+B$, A^2, A^{-1} are also symmetric positive definite. In each case, is the converse true? What can we say about the matrices BAB and ABA?

23.4. Let A and B be $n \times n$ symmetric positive definite matrices such that $AB = BA$. Show that AB is positive definite. Is the converse true?

23.5. Let S be the set of all $n \times n$ symmetric positive definite matrices. Is S a subspace of $M^{n \times n}$?

23.6. Let $A \in R^{n \times n}$ be a symmetric positive definite matrix. Show that for columns vectors the function $(u, v) = u^t A v$ is an inner product on R^n.

23.7. The matrix C in Problem 14.4 is a symmetric positive definite matrix.

23.8. Show that the matrix $A_n(2)$ given in (4.2) is a symmetric positive definite matrix.

23.9. Use Theorem 23.5 to show that the following symmetric matrices are positive definite

$$A = \begin{pmatrix} 1 & 2 & 3 \\ 2 & 8 & 12 \\ 3 & 12 & 34 \end{pmatrix}, \quad B = \begin{pmatrix} 16 & 4 & 4 & -4 \\ 4 & 10 & 4 & 2 \\ 4 & 4 & 6 & -2 \\ -4 & 2 & -2 & 4 \end{pmatrix}, \quad C = \begin{pmatrix} 2 & 1 & 1 & 2 \\ 1 & 3 & 2 & 1 \\ 1 & 2 & 4 & 3 \\ 2 & 1 & 3 & 5 \end{pmatrix}.$$

23.10. Use Theorem 23.5 to show that the following symmetric matrix is not positive definite

$$D = \begin{pmatrix} 2 & 1 & 1 & 2 & 1 & 2 \\ 1 & 3 & 2 & 1 & 3 & 4 \\ 1 & 2 & 4 & 3 & 1 & 0 \\ 2 & 1 & 3 & 5 & 1 & 2 \\ 1 & 3 & 1 & 1 & 0 & 1 \\ 2 & 4 & 0 & 2 & 1 & 5 \end{pmatrix}.$$

23.11. Find Cholesky decomposition for the symetric matrices $A, B,$ and C given in Problem 23.9.

23.12. For the quadratic form $q_3(x_1, x_2, x_3) = x_1^2 + 5x_2^2 + 2x_3^2 + 2\alpha x_1 x_2 + 2x_1 x_3 + 6x_2 x_3$, find the values of α so that it is positive definite.

Answers or Hints

23.1. If A is positive definite, $(e^k)^t A e^k = a_{kk} > 0,\ k = 1, \cdots, n.$ The converse is not true; consider the matrix $\begin{pmatrix} 2 & 5 \\ 5 & 3 \end{pmatrix}.$

23.2. Let $y = Cx \neq 0$ for $x \in R^n \backslash \{0\}$. Then, we have $x^t C^t A C x = y^t A y > 0,$ since A is positive definite.

23.3. If $x^t A x > 0$ and $x^t B x > 0$ for all $x \in R^n \backslash \{0\}$, then $x^t(A+B)x > 0,$ and hence $A + B$ is positive definite. The converse is not true, in fact, matrices $A = \begin{pmatrix} -1 & 0 \\ 0 & 2 \end{pmatrix}$ and $B = \begin{pmatrix} 3 & 0 \\ 0 & -1 \end{pmatrix}$ are not positive definite, but $C = A + B = \begin{pmatrix} 2 & 0 \\ 0 & 1 \end{pmatrix}$ is positive definite.

Since A is positive definite, from Theorem 23.2 it is nonsingular. Now from Problem 23.2 it follows that $A^2 = A^t I A$ is positive definite. The converse is not true. In fact, the matrix $A = \begin{pmatrix} -1 & 0 \\ 0 & 2 \end{pmatrix}$ is not positive definite, but $A^2 = \begin{pmatrix} 1 & 0 \\ 0 & 4 \end{pmatrix}$ is positive definite.

Let $x \in R^n \backslash \{0\}$ and $y = Ax \neq 0$. We have $y^t A^{-1} y = x^t A^t A^{-1} A x = x^t A x > 0.$ The converse follows by replacing A by $A^{-1}.$

Since matrices A and B are invertible, from Problem 23.2 it follows that matrices BAB and ABA are positive definite.

23.4. Since $AB = BA$ and matrices A and B are symmetric, AB is symmetric. Suppose λ is an eigenvalue of the matrix AB and $x \neq 0$ is the corresponding eigenvector. Then, $ABx = \lambda x$ implies that $x^t BABx = \lambda x^t Bx,$ and hence $\lambda = x^t BABx / x^t Bx > 0,$ since $x^t Bx > 0$ and $x^t BABx > 0.$ Thus, AB is positive definite. The converse is not true. For this, note that the matrix $C = \begin{pmatrix} 3 & 0 \\ 0 & 4 \end{pmatrix}$ is positive definite, and $C = AB = BA$ where

$A = \begin{pmatrix} -1 & 0 \\ 0 & 2 \end{pmatrix}$, $B = \begin{pmatrix} -3 & 0 \\ 0 & 2 \end{pmatrix}$. Clearly, the matrices A and B are not positive definite.

23.5. No, because a positive definite matrix multiplied by a negative scalar is not a positive definite matrix.

23.6. Because A is positive definite, $(u, u) = u^t A u > 0$. Because $u^t A v$ is a scalar, $(u^t A v)^t = u^t A v$. Also, $A^t = A$, because A is symmetric. Thus, $(u, v) = u^t A v = (u^t A v)^t = v^t A^t u^{tt} = v^t A u = (v, u)$. Finally, for any vectors u^1, u^2, v and scalars c_1, c_2, we have $(c_1 u^1 + c_2 u^2, v) = (c_1 u^{1t} + c_2 u^{2t}) A v = c_1 u^{1t} A v + c_2 u^{2t} A v = c_1 (u^1, v) + c_2 (u^2, v)$.

23.7. In view of Problem 14.4(i), the matrix C is symmetric. Now let u be any nonzero vector in R^n. Then, u will be same as the coordinates of some nonzero vector, say, v in V. Thus, from Problem 14.4(ii) it follows that $u^t C u = (u, u) > 0$.

23.8. From Problem 16.9, the eigenvalues of $A_n(2)$ are positive.

23.9. $\det(A^1) = 1, \det(A^2) = 4, \det(A^3) = 64$.
$\det(B^1) = 16, \det(B^2) = 144, \det(B^3) = 576, \det(B^4) = 576$.
$\det(C^1) = 2, \det(C^2) = 5, \det(C^3) = 13, \det(C^4) = 19$.

23.10. Follows from Problem 23.1. Also, note that $\det(D^1) = 2, \det(D^2) = 5, \det(D^3) = 13, \det(D^4) = 19, \det(D^5) = -72$.

23.11. $A = \begin{pmatrix} 1 & 0 & 0 \\ 2 & 2 & 0 \\ 3 & 3 & 4 \end{pmatrix} \begin{pmatrix} 1 & 2 & 3 \\ 0 & 2 & 3 \\ 0 & 0 & 4 \end{pmatrix}$.

$B = \begin{pmatrix} 4 & 0 & 0 & 0 \\ 1 & 3 & 0 & 0 \\ 1 & 1 & 2 & 0 \\ -1 & 1 & -1 & 1 \end{pmatrix} \begin{pmatrix} 4 & 1 & 1 & -1 \\ 0 & 3 & 1 & 1 \\ 0 & 0 & 2 & -1 \\ 0 & 0 & 0 & 1 \end{pmatrix}$.

$C = \begin{pmatrix} 1.4142 & 0.0000 & 0.0000 & 0.0000 \\ 0.7071 & 1.5811 & 0.0000 & 0.0000 \\ 0.7071 & 0.9487 & 1.6125 & 0.0000 \\ 1.4142 & 0.0000 & 1.2403 & 1.2089 \end{pmatrix} \begin{pmatrix} 1.4142 & 0.7071 & 0.7071 & 1.4142 \\ 0.0000 & 1.5811 & 0.9487 & 0.0000 \\ 0.0000 & 0.0000 & 1.6125 & 1.2403 \\ 0.0000 & 0.0000 & 0.0000 & 1.2089 \end{pmatrix}$.

23.12. The matrix of the quadratic form is $A = \begin{pmatrix} 1 & \alpha & 1 \\ \alpha & 5 & 3 \\ 1 & 3 & 2 \end{pmatrix}$. We apply Sylvester's criterion: $\det(A^1) = 1 > 0, \det(A^2) = 5 - \alpha^2 > 0$, which is true if $\alpha \in (-\sqrt{5}, \sqrt{5})$, and $\det(A^3) = (-2)(\alpha - 1)(\alpha - 2) > 0$, which is true if $\alpha \in (1, 2)$. Thus, $\alpha \in (1, 2)$.

Chapter 24

Moore–Penrose Inverse

In Chapter 4 we discussed the inverse of an $n \times n$ matrix. In this chapter we shall introduce the concept of a pseudo/generalized (Moore–Penrose) inverse, which is applicable to all $m \times n$ matrices. As an illustration we shall apply Moore–Penrose inverse to least squares solutions of linear equations.

From Remark 19.2 it follows that each $n \times n$ invertible matrix A has the reduced singular value decomposition

$$A = U_1 D_{nn} V_1^t; \tag{24.1}$$

where $U_1 = (u^1, \cdots, u^n)$, $V_1 = (v^1, \cdots, v^n)$ are $n \times n$ orthogonal matrices, and D_{nn} is the diagonal matrix with elements $\sigma_1 \geq \cdots \geq \sigma_n > 0$. Thus, from (24.1), we find

$$A^{-1} = V_1 D_{nn}^{-1} U_1^t. \tag{24.2}$$

Since the inverse of a matrix is unique, the right side of (24.2) provides another (factorized) representation (see (4.1)) of A^{-1}.

Following the lead of the representation (24.2), for any $m \times n$ matrix A with rank r, we define the Moore–Penrose inverse as

$$A^+ = V_1 D_{rr}^{-1} U_1^t; \tag{24.3}$$

here, V_1, D_{rr}^{-1}, U_1^t, respectively, are $n \times r, r \times r, r \times m$ matrices.

Example 24.1. For the matrix A in Example 4.2, (24.2) gives

$$A^{-1} \simeq \begin{pmatrix} -0.4296 & 0.8082 & 0.4028 \\ -0.4667 & -0.5806 & 0.6671 \\ 0.7731 & 0.0986 & 0.6266 \end{pmatrix} \begin{pmatrix} 0.1434 & 0.0000 & 0.0000 \\ 0.0000 & 0.3742 & 0.0000 \\ 0.0000 & 0.0000 & 0.6654 \end{pmatrix}$$
$$\times \begin{pmatrix} 0.0546 & -0.8732 & -0.4843 \\ 0.8590 & 0.2883 & -0.4231 \\ 0.5091 & -0.3929 & 0.7658 \end{pmatrix}.$$

Example 24.2. For the matrix A in Example 19.1, from Example 19.2

and (24.3) it follows that

$$A^+ = \begin{pmatrix} \frac{2}{3} & \frac{1}{\sqrt{3}} \\ \frac{2}{3} & -\frac{1}{\sqrt{3}} \\ \frac{1}{3} & 0 \\ 0 & \frac{1}{\sqrt{3}} \end{pmatrix} \begin{pmatrix} \frac{1}{3} & 0 \\ 0 & \frac{1}{\sqrt{3}} \end{pmatrix} \begin{pmatrix} 1 & 0 \\ 0 & 1 \end{pmatrix}.$$

A formal definition of Moore–Penrose inverse is as follows:

Definition 24.1. For a given $m \times n$ matrix A the $n \times m$ matrix A^+ is called Moore–Penrose inverse if the following hold:

(a) $AA^+A = A$, (b) $A^+AA^+ = A^+$, (c) $(AA^+)^t = AA^+$, (d) $(A^+A)^t = A^+A$.
(24.4)

From elementary calculations it follows that A^+ given in (24.3) satisfies all the four equations (24.4), and hence every matrix A has a Moore–Penrose inverse A^+. In fact, the equations in (24.4) determine A^+ uniquely. For this, let B^+ also be a Moore–Penrose inverse of A, i.e.,

(ã) $AB^+A = A$, (b̃) $B^+AB^+ = B^+$, (c̃) $(AB^+)^t = AB^+$, (d̃) $(B^+A)^t = B^+A$.
(24.5)

Now from the eight equations in (24.4) and (24.5), we have

$$
\begin{aligned}
A^+ &=^{(b)} A^+AA^+ =^{(d)} A^t(A^+)^tA^+ =^{(\tilde{a})} A^t(B^+)^tA^t(A^+)^tA^+ \\
&=^{(\tilde{d})} B^+AA^t(A^+)^tA^+ =^{(d)} B^+AA^+AA^+ =^{(a)} B^+AA^+,
\end{aligned}
$$

and similarly

$$
\begin{aligned}
B^+ &=^{(\tilde{b})} B^+AB^+ =^{(\tilde{c})} B^+(B^+)^tA^t =^{(a)} B^+(B^+)^tA^t(A^+)^tA^t \\
&=^{(c)} B^+(B^+)^tA^tAA^+ =^{(\tilde{c})} B^+AB^+AA^+ =^{(\tilde{a})} B^+AA^+.
\end{aligned}
$$

We summarize our above considerations in the following result.

Theorem 24.1. For any $m \times n$ matrix A the Moore–Penrose inverse A^+ exists uniquely.

From Example 11.4 it is clear that if the right inverse (and similarly the left inverse) of an $m \times n$ matrix exists then it may not be unique. In what follows, we shall show that the right and left inverses given in Problems 11.9 and 11.10 are in fact Moore–Penrose inverses, and hence unique. For this, we recall that an $m \times n$ matrix is said to have a *full row (column) rank* if and only if AA^t (A^tA) is invertible, which is equivalent to $r(A) = m$ $(r(A) = n)$.

Theorem 24.2. For an $m \times n$ matrix A the Moore–Penrose inverse is

$$A^+ = \begin{cases} A^t(AA^t)^{-1}, & \text{if } r(A) = m \\ (A^tA)^{-1}A^t, & \text{if } r(A) = n. \end{cases}$$
(24.6)

Proof. Assume that $r(A) = n$, then from the singular value decomposition it follows that

$$A^t A = (V_1 D_{nn}^t U_1^t)(U_1 D_{nn} V_1^t) = V_1 D_{nn}(U_1^t U_1) D_{nn} V_1^t = V_1 D_{nn}^2 V_1^t.$$

Since $r(A) = r(A^t A) = n$, the matrix $A^t A$ is invertible, and thus we have

$$(A^t A)^{-1} A^t = (V_1 D_{nn}^{-2} V_1^t)(V_1 D_{nn}^t U_1^t)$$
$$= V_1 D_{nn}^{-2}(V_1^t V_1) D_{nn} U_1^t = V_1 D_{nn}^{-1} U_1^t = A^+. \quad \blacksquare$$

Example 24.3. For the matrix A in (11.2), we have $r(A) = m = 3$, thus the Moore–Penrose inverse exists uniquely and simple calculations give

$$A^+ = A^t(AA^t)^{-1} = \begin{pmatrix} \frac{25}{14} & -\frac{3}{7} & -\frac{4}{7} \\ -6 & 3 & 1 \\ \frac{23}{14} & -\frac{5}{7} & -\frac{2}{7} \\ \frac{11}{7} & -\frac{6}{7} & -\frac{1}{7} \end{pmatrix},$$

which is the same as (11.4) with $a = 11/7, b = -6/7, c = -1/7$. For other choices of a, b, c at least one of the four conditions in Definition 24.1 will fail; for example, if we choose $a = b = c = 2$, then the condition (d), i.e., $(A^+ A)^t = A^+ A$, does not hold.

Example 24.4. For the matrix A in Example 19.1, we have $r(A) = m = 2$, thus the Moore–Penrose inverse exists uniquely and simple calculations give

$$A^+ = A^t(AA^t)^{-1} = \begin{pmatrix} \frac{2}{9} & \frac{1}{3} \\ \frac{2}{9} & -\frac{1}{3} \\ \frac{1}{9} & 0 \\ 0 & \frac{1}{3} \end{pmatrix}.$$

Our next result provides some basic properties of Moore–Penrose inverse.

Theorem 24.3. Let the $m \times n$ matrix A have the rank r, and let A^+ be its Moore–Penrose inverse given in (24.3). Then, the following hold:

(i) for every vector $y \in R^m$ the vector $A^+ y \in R(A)$

(ii) for every vector $y \in \mathcal{N}(A)$ the vector $A^+ y = 0$

(iii) $A^+ u^i = v^i/\sigma_i,\ i = 1, \cdots, r.$

Proof. (i) Since $A^+ y = (V_1 D_{rr}^{-1} U_1^t)y = V_1(D_{rr}^{-1} U_1^t y)$, it follows that $A^+ y$ is a linear combination of the columns of V_1. But this in view of Theorem 19.2(iii) implies that $A^+ y \in R(A)$.

(ii) If $y \in \mathcal{N}(A)$, then clearly $y \perp C(A)$, which in view of Theorem 19.2(i) implies that $y \perp U_1$. But this in view of Problem 14.11 implies that $U_1^t y = 0$, and hence, we have $A^+ y = (V_1 D_{rr}^{-1}) U_1^t y = 0$.

(iii)   From (24.3), we have $A^+U_1 = V_1 D_{rr}^{-1}$, which on comparing the vectors on both sides immediately gives $A^+u^i = v^i/\sigma_i$, $i = 1, \cdots, r$.   ∎

Now we shall apply Moore–Penrose inverse to least squares solutions of the $m \times n$ linear system (5.2). For this, we recall that Theorem 21.1 assures the existence of a unique least squares solution (21.3) provided A is a full column matrix, i.e., $r(A) = n$. This unique solution in view of Theorem 24.2 can be written as

$$\hat{x} = A^+b. \tag{24.7}$$

Theorem 21.1 also says that if the columns of A are linearly dependent, i.e., $r(A) \neq n$, then there are an infinite number of least squares solutions. However, since for each $m \times n$ matrix A the Moore–Penrose inverse A^+ exists uniquely, the representation (24.7) is meaningful and provides $\hat{x}$ uniquely. Further, in view of Theorem 24.3(i), $A^+b \in R(A)$. In the following result, we shall show that (24.7) is, in fact, the unique least squares solution of the system (5.2) even when $r(A) \neq n$.

Theorem 24.4.   For each $m \times n$ matrix A and $b \in R^m$, the system (5.2) has a unique least squares solution given by (24.7).

Proof.   From our above discussion and Theorem 21.1 it suffices to show that $\hat{x} = A^+b$ satisfies the normal equations (21.2), i.e., $A^t A\hat{x} = A^t b$. For this, from (19.2) and (24.3), we have

$$
\begin{aligned}
(A^t A)A^+b &= (V_1 D_{rr} U_1^t U_1 D_{rr} V_1^t)(V_1 D_{rr}^{-1} U_1^t b) \\
&= V_1 D_{rr}(U_1^t U_1)D_{rr}(V_1^t V_1)D_{rr}^{-1}U_1^t b \\
&= V_1 D_{rr} U_1^t b = A^t b,
\end{aligned}
$$

i.e., A^+b is a solution of the normal equations (21.2).   ∎

Example 24.5.   Consider the system

$$A(x_1, x_2, x_3, x_4)^t = (b_1, b_2, b_3)^t, \tag{24.8}$$

which has an infinite number of solutions

$$
\begin{pmatrix} x_1 \\ x_2 \\ x_3 \\ x_4 \end{pmatrix} = \begin{pmatrix} b_1 - \frac{1}{2}b_3 \\ -6b_1 + 3b_2 + b_3 \\ 4b_1 - 2b_2 - \frac{1}{2}b_3 \\ 0 \end{pmatrix} + \begin{pmatrix} \frac{1}{2} \\ 0 \\ -\frac{3}{2} \\ 1 \end{pmatrix} c, \tag{24.9}
$$

where c is an arbitrary constant. We also note that from Theorems 24.2 and 24.4 and Example 24.3 the unique least squares solution of (24.8) is

$$
\hat{x} = A^+(b_1, b_2, b_3) = \begin{pmatrix} \frac{25}{14}b_1 - \frac{3}{7}b_2 - \frac{4}{7}b_3 \\ -6b_1 + 3b_2 + b_3 \\ \frac{23}{14}b_1 - \frac{5}{7}b_2 - \frac{2}{7}b_3 \\ \frac{11}{7}b_1 - \frac{6}{7}b_2 - \frac{1}{7}b_3 \end{pmatrix}. \tag{24.10}
$$

Clearly, when $c = \frac{11}{7}b_1 - \frac{6}{7}b_2 - \frac{1}{7}b_3$, (24.9) becomes the same as (24.10).

Example 24.6. For the matrix A in Example 21.2, $r(A) = 2$. The reduced singular value decomposition of this matrix appears as

$$A = U_1 D_{rr} V_1^t = \begin{pmatrix} 0.4968 & -0.0293 \\ -0.4692 & 0.8216 \\ 0.5874 & 0.5325 \\ 0.4337 & 0.2013 \end{pmatrix} \begin{pmatrix} 7.5246 & 0.0000 \\ 0.0000 & 5.7776 \end{pmatrix}$$

$$\times \begin{pmatrix} 0.3065 & 0.7812 & 0.5438 \\ 0.8599 & -0.4722 & 0.1938 \end{pmatrix}.$$

Thus, the Moore–Penrose inverse is

$$A^+ = V_1 D_{rr}^{-1} U_1^t = \begin{pmatrix} 0.0159 & 0.1032 & 0.1031 & 0.0476 \\ 0.0540 & -0.1158 & 0.0175 & 0.0286 \\ 0.0349 & -0.0064 & 0.0603 & 0.0381 \end{pmatrix}.$$

From this, we can directly compute the unique least squares solution of (21.12) as

$$\hat{x} = A^+ b = \begin{pmatrix} 0.7220 \\ -0.0107 \\ 0.3554 \end{pmatrix}.$$

This shows that in (21.14), $c \simeq 0.3554$.

Problems

24.1. Show that (i) $(A^t)^+ = (A^+)^t$, (ii) $A^{++} = A$.

24.2. Show that if in Definition 24.1, A^+ satisfies only equation (a), then it is not unique.

24.3. Give an example to show that $(AB)^+ \neq B^+ A^+$.

24.4. Let the $m \times n$ matrix A have the rank r, and let A^+ be its Moore–Penrose inverse given in (24.3). Show that

(i) AA^+ is the orthogonal projection of R^m onto $C(A)$
(ii) $A^+ A$ is the orthogonal projection of R^n onto $R(A)$.

24.5. Find the Moore–Penrose inverse of the matrices given in Problem 4.1.

24.6. Find the Moore–Penrose inverse of the matrices given in Problems 19.1–19.4.

24.7. Find the unique least squares solutions of Problems 21.1–21.5.

Answers or Hints

24.1. (i) From (19.2) and (24.3), we have $A = U_1 D_{rr} V_1^t$ and $A^+ = V_1 D_{rr}^{-1} U_1^t$.
Thus, $A^t = V_1 D_{rr} U_1^t$ and $(A^t)^+ = U_1 D_{rr}^{-1} V_1^t = (A^+)^t$.
(ii) Since $A^+ = V_1 D_{rr}^{-1} U_1^t$, we have $A^{++} = U_1 (D_{rr}^{-1})^{-1} V_1^t = U_1 D_{rr} V_1^t = A$.

24.2. For every $n \times m$ matrix P, the matrix $B = A^+ + (P - A^+ A P A A^+)$ satisfies (a).

24.3. For the matrices $A = (5, 2)$, $B = \begin{pmatrix} 1 \\ 3 \end{pmatrix}$, we have $A^+ \simeq \begin{pmatrix} 0.1724 \\ 0.0690 \end{pmatrix}$,
$B^+ \simeq (0.1, 0.3)$, $(AB)^+ \simeq 0.0909$, $B^+ A^+ \simeq 0.0379$, and hence $(AB)^+ \neq B^+ A^+$.

24.4. We need to show that $R(A) \oplus N(A^+) = R^m$ and $R(A^+) \oplus N(A) = R^n$.
(i) Since $A = U_1 D_{rr} V_1^t$ and $A^+ = V_1 D_{rr}^{-1} U_1^t$ for $y \in R^m$, we have $AA^+ y = U_1(D_{rr} V_1^t V_1 D_{rr}^{-1} U_1^t y)$. This means $AA^+ y$ is a linear combination of the rows of U_1, and thus from Theorem 19.2(i), we have $AA^+ y \in C(A)$.
(ii) For $y \in R^n$, we have $A^+ A y = V_1(D_{rr}^{-1} U_1^t U_1 D_{rr} V_1^t y)$, and hence $A^+ A y$ is a linear combination of the rows of V_1, and now from Theorem 19.2(iii), we have $A^+ A y \in R(A)$.

24.5. $A^+ = \begin{pmatrix} 0.2222 & 1.0556 & -0.6111 \\ -0.1111 & -0.7778 & 0.5556 \\ 0.1111 & 0.2778 & -0.0556 \end{pmatrix}$.

$B^+ = \begin{pmatrix} \frac{7}{2} & -2 & \frac{1}{2} \\ -\frac{13}{4} & \frac{11}{4} & -\frac{3}{4} \\ \frac{3}{4} & -\frac{3}{4} & \frac{1}{4} \end{pmatrix}$.

$C^+ = \begin{pmatrix} 0.0256 & 0.0858 & -0.1925 \\ -0.0009 & -0.0580 & 0.1229 \\ 0.0483 & -0.0025 & -0.0162 \end{pmatrix}$.

24.6. $\begin{pmatrix} 1 & 2 & 1 & 0 \\ 2 & 0 & 1 & 1 \end{pmatrix}^+ = \begin{pmatrix} 0 & \frac{1}{3} \\ \frac{4}{9} & -\frac{2}{9} \\ \frac{1}{9} & \frac{1}{9} \\ -\frac{1}{9} & \frac{2}{9} \end{pmatrix}$.

$\begin{pmatrix} 1 & 3 \\ -3 & 3 \\ -3 & 1 \\ 1 & 1 \end{pmatrix}^+ = \frac{1}{84}\begin{pmatrix} 11 & -9 & -13 & 7 \\ 17 & 9 & -1 & 7 \end{pmatrix}$.

$\begin{pmatrix} -1 & 0 & 1 \\ -1 & 1 & 2 \\ 0 & 1 & 1 \end{pmatrix}^+ = \frac{1}{9}\begin{pmatrix} -5 & -1 & 4 \\ -4 & 1 & 5 \\ 1 & 2 & 1 \end{pmatrix}$.

$\begin{pmatrix} 1 & 2 & 3 \\ 2 & 1 & 0 \\ 1 & 1 & 2 \\ 0 & 3 & 4 \end{pmatrix}^+ = \begin{pmatrix} 0.1818 & 0.2273 & 0.3636 & -0.3182 \\ -0.2909 & 0.5364 & -0.7818 & 0.6091 \\ 0.2727 & -0.4091 & 0.5455 & -0.2273 \end{pmatrix}$.

24.7. Problem 21.1, $(0.3333, 0.6667, 1.0000)^t$.
Problem 21.2, $(0.1759, 0.5548, 0.0114, 0.3648)^t$.
Problem 21.3, $(0.1990, 0.3026, 0.2828, -0.4877, -0.3693)^t$.
Problem 21.4, $(0.2020, 0.4039, -0.2824, 0.0412)^t$.
Problem 21.5, $(0.5136, 0.7113, -0.1810)^t$.

Chapter 25

Special Matrices

In this last chapter we shall briefly discuss irreducible, nonnegative, diagonally dominant, monotone, and Toeplitz matrices. These matrices frequently occur in numerical solutions of differential and integral equations, spline functions, problems and methods in physics, statistics, signal processing, discrete Fourier transform, and the study of cyclic codes for error correction.

An $n \times n$ matrix $A = (a_{ij})$ is said to be *reducible* if the set of indices $N = \{1, 2, \cdots, n\}$ can be divided into two nonempty disjoint sets S and T with $N = S \cup T$ such that $a_{ij} = 0$ for all $i \in S$ and $j \in T$. A square matrix is called *irreducible* if it is not reducible.

Example 25.1. The following matrices are reducible

$$A = \begin{pmatrix} 5 & 3 & 0 \\ 6 & 4 & 0 \\ 0 & 6 & 9 \end{pmatrix}, \quad B = \begin{pmatrix} 3 & 2 & 5 \\ 0 & 7 & 1 \\ 0 & 4 & 3 \end{pmatrix}.$$

Clearly, for the matrix A, if we take $S = \{1, 2\}$, $T = \{3\}$ then $a_{13} = a_{23} = 0$, whereas for the matrix B if $S = \{2, 3\}$, $T = \{1\}$ then $a_{21} = a_{31} = 0$. The following matrices are irreducible

$$C = \begin{pmatrix} 5 & 3 & 1 \\ 6 & 4 & 0 \\ 0 & 6 & 9 \end{pmatrix}, \quad D = \begin{pmatrix} 3 & 2 & 5 \\ 0 & 7 & 1 \\ 1 & 4 & 3 \end{pmatrix}.$$

For this, we need to consider all six possible particians of the set $\{1, 2, 3\}$, i.e., $P_1 : S = \{1\}, T = \{2, 3\}$; $P_2 : S = \{2\}, T = \{1, 3\}$; $P_3 : S = \{3\}, T = \{1, 2\}$; $P_4 : S = \{1, 2\}, T = \{3\}$; $P_5 : S = \{2, 3\}, T = \{1\}$; and $P_6 : S = \{1, 3\}, T = \{2\}$. For the matrix C in each of these particians, we have $a_{12} = 3, a_{21} = 6, a_{32} = 6, a_{13} = 1, a_{21} = 6, a_{12} = 3$. Similarly, for the matrix D, we have $a_{12} = 2, a_{23} = 1, a_{32} = 4, a_{13} = 5, a_{31} = 1, a_{12} = 2$.

The following result provides necessary and sufficient conditions for an $n \times n$ matrix to be reducible.

Theorem 25.1. An $n \times n$ matrix A is reducible if and only if there exists a permutation matrix P such that

$$P^t A P = \begin{pmatrix} A_{11} & 0 \\ A_{21} & A_{22} \end{pmatrix};$$

213

here A_{11} and A_{22} are square matrices of orders r and $n - r$, respectively, A_{21} is an $(n - r) \times r$ matrix, and 0 is the $r \times (n - r)$ null matrix, $1 \leq r \leq n - 1$.

Example 25.2. In view of Theorem 25.1, matrix A in Example 25.1 is reducible. For the matrix

$$A = \begin{pmatrix} 4 & 3 & 5 & 1 \\ 0 & 2 & 0 & 3 \\ 3 & 5 & 1 & 2 \\ 0 & 2 & 0 & 1 \end{pmatrix}$$

the permutation matrix $P = (e^4, e^2, e^1, e^3)$ gives

$$P^t A P = \begin{pmatrix} 1 & 2 & 0 & 0 \\ 3 & 2 & 0 & 0 \\ 1 & 3 & 4 & 5 \\ 2 & 5 & 3 & 1 \end{pmatrix} = B. \tag{25.1}$$

Thus, from Theorem 25.1 it follows that the matrix A is reducible.

Let $v_1, \cdots, v_n$ be n distinct points in the xy–plane. For each $a_{ij} \neq 0$, $1 \leq i, j \leq n$ we connect the points v_i with v_j with line segments directed from v_i to v_j. In *graph theory* the points $v_1, \cdots, v_n$ are called vertices, nodes, or points, and the line segments are called edges, arcs, or simply lines. The graph so constructed is called a *directed graph*, because edges are directed from one vertex to another. A graph is called *strongly connected* if for any ordered pair of nodes v_i, v_j, there exists a path $v_i v_{k_1} \to v_{k_1} v_{k_2} \to \cdots \to v_{k_m} v_j$ connecting v_i to v_j.

Theorem 25.2. An $n \times n$ matrix $A = (a_{ij})$ is irreducible if and only if its directed graph is strongly connected.

Example 25.3. Directed graphs of matrices A and C in Example 25.1 appear as

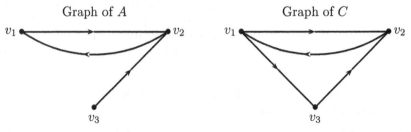

Clearly, the graph of A is not strongly connected, whereas of C is strongly connected.

From Theorem 25.2 the following corollary is immediate.

Corollary 25.1. An $n \times n$ tridiagonal matrix $A = (a_{ij})$ is irreducible if and only if $a_{i,i+1} \neq 0$, $i = 1, \cdots, n-1$ and $a_{i,i-1} \neq 0$, $i = 2, \cdots, n$.

Matrices in (4.2) and (4.16) are irreducible.

An $m \times n$ matrix $A = (a_{ij})$ is said to be *nonnegative* (*positive*) if $a_{ij} \geq (>) 0$, $1 \leq i \leq m, 1 \leq j \leq n$. Eigenvalues and eigenvectors of nonnegative irreducible $n \times n$ matrices are described by the following result.

Theorem 25.3 (Perron–Frobenius). If A is an $n \times n$, nonnegative irreducible matrix, then the following hold:

(i)   one of its eigenvalues, say, λ^* is positive, and if λ is any other eigenvalue, then $|\lambda| \leq \lambda^*$

(ii) there is a positive eigenvector v^* corresponding to the eigenvalue λ^*

(iii) the eigenvalues of modulus λ^* are simple

(iv) all eigenvalues of modulus λ^* are of the form

$$\lambda_k = \lambda^* \exp\left(\frac{2\pi k \sqrt{-1}}{m}\right), \quad k = 0, 1, \cdots, m-1.$$

Example 25.4. For the matrix C in Example 25.1, $\lambda^* \simeq 10.740084$, $v^* \simeq (1, 0.890197, 3.069495)^t$, and for the matrix D, $\lambda^* \simeq 8.074555$, $v^* = (1, 0.688283, 0.739598)^t$.

An $n \times n$ matrix $A = (a_{ij})$ is said to be *diagonally dominant* if for every row of the matrix, the magnitude of the diagonal entry in a row is larger than or equal to the sum of the magnitudes of all the other (non-diagonal) entries in that row, i.e.,

$$|a_{ii}| \geq \sum_{j=1,j\neq i}^{n} |a_{ij}| \quad \text{for all} \quad 1 \leq i \leq n. \tag{25.2}$$

If in (25.2) a strict inequality holds, then A is called *strictly diagonally dominant*.

Example 25.5. None of the matrices A, B, C, D in Example 25.1 are diagonally dominant. Matrix $A_n(x)$ defined in (4.2) is diagonally dominant if $|x| = 2$, and strictly diagonally dominant if $|x| > 2$. Consider the matrices

$$A = \begin{pmatrix} 5 & 3 & 0 \\ 6 & 8 & 0 \\ 0 & 6 & 9 \end{pmatrix}, \quad B = \begin{pmatrix} 5 & 3 & 2 \\ 8 & 8 & 0 \\ 0 & 9 & 9 \end{pmatrix}.$$

The matrix A is strictly diagonally dominant but not irreducible, whereas matrix B is diagonally dominant and irreducible. In view of Corollary 25.1, if

$|x| = 2$, then the matrix $A_n(x)$ in (4.2) is diagonally dominant and irreducible, and if $|x| > 2$, then it is strictly diagonally dominant and irreducible.

Theorem 25.4. If an $n \times n$ matrix A is strictly diagonally dominant, or diagonally dominant and irreducible, then A is invertible.

Example 25.6. From Theorem 25.4 it is clear that the matrix $A_n(x)$ in (4.2) for $|x| \geq 2$ is invertible. In particular, $A_5(3)$ is invertible; see Example 4.3. Matrices A and B in Example 25.5 are invertible and their inverses appear as

$$
A^{-1} = \begin{pmatrix} \frac{4}{11} & -\frac{3}{22} & 0 \\ -\frac{3}{11} & \frac{5}{22} & 0 \\ \frac{2}{11} & -\frac{5}{33} & \frac{1}{9} \end{pmatrix}, \quad
B^{-1} = \begin{pmatrix} \frac{1}{4} & -\frac{1}{32} & -\frac{1}{18} \\ -\frac{1}{4} & \frac{5}{32} & \frac{1}{18} \\ \frac{1}{4} & -\frac{5}{32} & \frac{1}{18} \end{pmatrix}.
$$

The converse of Theorem 25.4 does not hold. For this, we note that the matrices A and B in Example 25.1 are neither diagonally dominant nor irreducible, but their inverse exists:

$$
A^{-1} = \begin{pmatrix} 2 & -\frac{3}{2} & 0 \\ -3 & \frac{5}{2} & 0 \\ 2 & -\frac{5}{3} & \frac{1}{9} \end{pmatrix}, \quad
B^{-1} = \begin{pmatrix} \frac{1}{3} & \frac{14}{51} & -\frac{11}{17} \\ 0 & \frac{3}{17} & -\frac{1}{17} \\ 0 & -\frac{4}{17} & \frac{7}{17} \end{pmatrix}.
$$

Let $A = (a_{ij})$ be an $m \times n$ matrix. In what follows by $A \geq 0$, we mean that the matrix A is nonnegative. An $n \times n$ matrix A is said to be *monotone* if $Au \geq 0$ implies that u is nonnegative, i.e., $u = (u_1, \cdots, u_n)^t \geq 0$.

If a matrix A is monotone, and $Au \leq 0$, then it follows that $-Au \geq 0$, which implies that $A(-u) \geq 0$, and hence $-u \geq 0$, or $u \leq 0$. Thus, if A is monotone, and $Au = 0$, then u must simultaneously satisfy $u \geq 0$ and $u \leq 0$, i.e., $u = 0$. This simple observation leads to the following result.

Theorem 25.5. If an $n \times n$ matrix A is monotone, then $\det(A) \neq 0$, i.e, it is nonsingular.

Example 25.7. The converse of Theorem 25.5 does not hold. For this, it suffices to note that for the matrix B in Example 25.1, $\det(B) = 51$ and $B(1, 1, -1)^t = (0, 6, 1)$.

The following result provides necessary and sufficient conditions so that an $n \times n$ matrix A is monotone.

Theorem 25.6. An $n \times n$ matrix A is monotone if and only if $A^{-1} \geq 0$.

Example 25.8. From Theorem 25.6 and Example 25.6 it follows that matrices A and B in Examples 25.1 are 25.5 are not monotone. Thus, reducible, irreducible, and diagonally dominant matrices are not necessarily monotone. In view of Example 4.3, the matrix $A_5(3)$ is monotone.

The following easily verifiable results provide sufficient conditions for an $n \times n$ matrix A to be monotone.

Theorem 25.7. Let (in addition to irreducibility) an $n \times n$ matrix $A = (a_{ij})$ satisfy the following conditions:

(i) $a_{ij} \leq 0$, $i \neq j$, $1 \leq i, j \leq n$

(ii) $\sum_{j=1}^{n} a_{ij} \geq 0$, $1 \leq i \leq n$ with strict inequality for at least one i.

Then, the matrix A is monotone.

Example 25.9. In view of Corollary 25.1, the matrix $A_n(x)$ for $x \geq 2$ satisfies all conditions of Theorem 25.7. Consider the matrices

$$A = \begin{pmatrix} 5 & -3 & -2 \\ -8 & 8 & 0 \\ 0 & -9 & 10 \end{pmatrix}, \quad B = \begin{pmatrix} -5 & 3 & 3 \\ 8 & -8 & 0 \\ 0 & 9 & -9 \end{pmatrix}, \quad C = \begin{pmatrix} 5 & -3 & 0 \\ -6 & 4 & 0 \\ 0 & -6 & 9 \end{pmatrix}$$

and their inverses

$$A^{-1} = \begin{pmatrix} 5 & 3 & 1 \\ 5 & \frac{25}{8} & 1 \\ \frac{9}{2} & \frac{45}{16} & 1 \end{pmatrix}, \quad B^{-1} = \begin{pmatrix} 1 & \frac{3}{4} & \frac{1}{3} \\ 1 & \frac{5}{8} & \frac{1}{3} \\ 1 & \frac{5}{8} & \frac{2}{9} \end{pmatrix}, \quad C^{-1} = \begin{pmatrix} 2 & \frac{3}{2} & 0 \\ 3 & \frac{5}{2} & 0 \\ 2 & \frac{5}{3} & \frac{1}{9} \end{pmatrix}.$$

From Theorem 25.6 all three matrices A, B, and C are monotone, but only A satisfies all conditions of Theorem 25.7.

Remark 25.1. From the matrices in Example 25.9, we note that $A + B$ is a singular matrix, and hence Theorem 25.5 implies that the addition of two monotone matrices may not be monotone. In view of Theorem 25.6, matrix A^{-1} is not monotone, and hence the inverse of a monotone matrix may not be monotone. However, from $(AB)^{-1} = B^{-1}A^{-1}$ it is clear that if A and B are monotone, then both AB and BA are also monotone.

Theorem 25.8. Let the $n \times n$ matrices A and C be monotone, and the matrix B be such that $A \leq B \leq C$. Then, the matrix B is also monotone.

Example 25.10. Clearly, the following matrices

$$A = \begin{pmatrix} 1 & -1 & -1 \\ -2 & 3 & 1 \\ -1 & 1 & 2 \end{pmatrix}, \quad B = \begin{pmatrix} 1 & -1 & -1 \\ -2 & 4 & 1 \\ -1 & 1 & 3 \end{pmatrix}, \quad C = \begin{pmatrix} 1 & -1 & -1 \\ -1 & 5 & 1 \\ -1 & 1 & 5 \end{pmatrix}$$

satisfy $A \leq B \leq C$, and since

$$A^{-1} = \begin{pmatrix} 5 & 1 & 2 \\ 3 & 1 & 1 \\ 1 & 0 & 1 \end{pmatrix} \quad \text{and} \quad C^{-1} = \begin{pmatrix} \frac{3}{2} & \frac{1}{4} & \frac{1}{4} \\ \frac{1}{4} & \frac{1}{4} & 0 \\ \frac{1}{4} & 0 & \frac{1}{4} \end{pmatrix}$$

matrices A and C are monotone. Thus, Theorem 25.8 is applicable, and the matrix B must be monotone, indeed we have

$$B^{-1} = \begin{pmatrix} \frac{11}{4} & \frac{1}{2} & \frac{3}{4} \\ \frac{5}{4} & \frac{1}{2} & \frac{1}{4} \\ \frac{1}{2} & 0 & \frac{1}{2} \end{pmatrix}.$$

Theorem 25.9. Let the $n \times n$ matrix A be written as $A = I - B$, where $B = (b_{ij}) \geq 0$ and (in any norm) $\|B\| < 1$. Then, the matrix A is monotone.

Example 25.11. Consider the following matrix and its inverse

$$A = \begin{pmatrix} 1 & -\frac{1}{2} & -\frac{1}{4} \\ -\frac{1}{4} & \frac{1}{2} & -\frac{1}{6} \\ -\frac{1}{6} & -\frac{1}{3} & 1 \end{pmatrix}, \quad A^{-1} = \begin{pmatrix} \frac{32}{19} & \frac{42}{19} & \frac{15}{19} \\ \frac{20}{19} & \frac{69}{19} & \frac{33}{38} \\ \frac{12}{19} & \frac{30}{19} & \frac{27}{19} \end{pmatrix}.$$

From Theorem 26.6 it follows that the matrix A is monotone. Since

$$A = I - B = \begin{pmatrix} 1 & 0 & 0 \\ 0 & 1 & 0 \\ 0 & 0 & 1 \end{pmatrix} - \begin{pmatrix} 0 & \frac{1}{2} & \frac{1}{4} \\ \frac{1}{4} & \frac{1}{2} & \frac{1}{6} \\ \frac{1}{6} & \frac{1}{3} & 0 \end{pmatrix}$$

the matrix B satisfies all conditions of Theorem 25.9, and hence matrix A is monotone.

Theorem 25.10. Let the $n \times n$ matrix A be symmetric, positive definite, and written as $A = I - B$, where $B = (b_{ij}) \geq 0$. Then, the matrix A is monotone.

Example 25.12. For the matrix

$$A = \begin{pmatrix} 1 & -\frac{1}{3} & 0 & 0 \\ -\frac{1}{3} & 1 & -\frac{1}{3} & 0 \\ 0 & -\frac{1}{3} & 1 & -\frac{1}{3} \\ 0 & 0 & -\frac{1}{3} & 1 \end{pmatrix}$$

eigenvalues are

$$\frac{5}{6} - \frac{1}{6}\sqrt{5}, \quad \frac{1}{6}\sqrt{5} + \frac{5}{6}, \quad \frac{7}{6} - \frac{1}{6}\sqrt{5}, \quad \frac{1}{6}\sqrt{5} + \frac{7}{6}$$

positive, and thus it is positive definite. We can write $A = I - B$, where

$$B = \begin{pmatrix} 0 & \frac{1}{3} & 0 & 0 \\ \frac{1}{3} & 0 & \frac{1}{3} & 0 \\ 0 & \frac{1}{3} & 0 & \frac{1}{3} \\ 0 & 0 & \frac{1}{3} & 0 \end{pmatrix}.$$

Since $\|B\|_\infty < 1$, conditions of Theorem 25.10 are satisfied, and thus the matrix A is monotone. Indeed, we find

$$A^{-1} = \frac{1}{55} \begin{pmatrix} 63 & 24 & 9 & 3 \\ 24 & 72 & 27 & 9 \\ 9 & 27 & 72 & 24 \\ 3 & 9 & 24 & 63 \end{pmatrix}.$$

An $m \times n$ matrix $A = (a_{ij})$ is called a Toeplitz matrix if $a_{ij} = a_{i+1,j+1} = a_{i-j}$. An $n \times n$ *Toeplitz matrix* has the form

$$A = \begin{pmatrix} a_0 & a_{-1} & a_{-2} & \cdots & \cdots & a_{-(n-1)} \\ a_1 & a_0 & a_{-1} & \cdots & \cdots & a_{-(n-2)} \\ a_2 & a_1 & a_0 & \cdots & \cdots & a_{-(n-3)} \\ \cdots & \cdots & \cdots & \cdots & \cdots & \cdots \\ a_{n-2} & \cdots & \cdots & a_1 & a_0 & a_{-1} \\ a_{n-1} & \cdots & \cdots & a_2 & a_1 & a_0 \end{pmatrix}.$$

In the above matrix A all diagonal elements are equal to a_0. Further, we note that this matrix has only $2n - 1$ degrees of freedom compared to n^2, thus it is easier to solve the systems $Ax = b$. For this, Levinson's algorithm is well known. An $n \times n$ Toeplitz matrix $A = (a_{ij})$ is called *symmetric* provided $a_{ij} = b_{|i-j|}$.

Example 25.13. For $n = 4$, Toeplitz and symmetric Toeplitz matrices, respectively, appear as

$$A = \begin{pmatrix} a_0 & a_{-1} & a_{-2} & a_{-3} \\ a_1 & a_0 & a_{-1} & a_{-2} \\ a_2 & a_1 & a_0 & a_{-1} \\ a_3 & a_2 & a_1 & a_0 \end{pmatrix}, \quad B = \begin{pmatrix} b_0 & b_1 & b_2 & b_3 \\ b_1 & b_0 & b_1 & b_2 \\ b_2 & b_1 & b_0 & b_1 \\ b_3 & b_2 & b_1 & b_0 \end{pmatrix}.$$

A symmetric Toeplitz matrix B is said to be *banded* if there is an integer $d < n-1$ such that $b_\ell = 0$ if $\ell \geq d$. In this case, we say that B has *bandwidth d*. Thus, an $n \times n$ banded symmetric Toeplitz matrix with bandwidth 2 appears as

$$B = \begin{pmatrix} b_0 & b_1 & & & & \\ b_1 & b_0 & b_1 & & & \\ & b_1 & b_0 & b_1 & & \\ & & \cdots & \cdots & & \\ & & & b_1 & b_0 & b_1 \\ & & & & b_1 & b_0 \end{pmatrix}. \tag{25.3}$$

Clearly, matrices (4.2) and (4.16) are symmetric Toeplitz matrices with bandwidth 2. For the matrix B in (25.3), following as in Problem 16.9, we find that the eigenvalues and the corresponding eigenvectors are

$$\lambda_i = b_0 + 2b_1 \cos\left(\frac{i\pi}{n+1}\right), \quad 1 \leq i \leq n$$

and

$$u^i = \left(\sin \frac{i\pi}{n+1}, \sin \frac{2i\pi}{n+1}, \cdots, \sin \frac{ni\pi}{n+1} \right)^t, \quad 1 \le i \le n,$$

also

$$\det(B) = \prod_{i=1}^{n} \left[b_0 + 2b_1 \cos \left(\frac{i\pi}{n+1} \right) \right].$$

Example 25.14. For the matrix B in (25.3) with $n = 4$, it follows that

$$\lambda_1 = b_0 + \frac{1+\sqrt{5}}{2} b_1, \; \lambda_2 = b_0 + \frac{1-\sqrt{5}}{2} b_1, \; \lambda_3 = b_0 - \frac{1-\sqrt{5}}{2} b_1, \; \lambda_4 = b_0 - \frac{1+\sqrt{5}}{2} b_1,$$

$$u^1 = \begin{pmatrix} 1 \\ \frac{1+\sqrt{5}}{2} \\ \frac{1+\sqrt{5}}{2} \\ 1 \end{pmatrix}, \; u^2 = \begin{pmatrix} 1 \\ \frac{1-\sqrt{5}}{2} \\ \frac{1-\sqrt{5}}{2} \\ 1 \end{pmatrix}, \; u^3 = \begin{pmatrix} -1 \\ \frac{1-\sqrt{5}}{2} \\ \frac{-1+\sqrt{5}}{2} \\ 1 \end{pmatrix}, \; u^4 = \begin{pmatrix} -1 \\ \frac{1+\sqrt{5}}{2} \\ -\frac{1+\sqrt{5}}{2} \\ 1 \end{pmatrix},$$

$$\det(B) = (b_1^2 - b_0^2 - b_0 b_1)(b_0 b_1 - b_0^2 + b_1^2).$$

In Toeplitz matrix A, if we take $a_i = a_{-(n-i)}, \; 1 = 1, \cdots, n-1$, then it reduces to a circulant matrix (see Problem 16.8),

$$A = \text{circ}(a_0, a_1, \cdots, a_{-(n-1)})$$

$$= \begin{pmatrix} a_0 & a_{-1} & a_{-2} & \cdots & & a_{-(n-1)} \\ a_{-(n-1)} & a_0 & a_{-1} & \cdots & & a_{-(n-2)} \\ a_{-(n-2)} & a_{-(n-1)} & a_0 & \cdots & & a_{-(n-3)} \\ \cdots & \cdots & \cdots & \cdots & \cdots & \cdots \\ a_{-2} & \cdots & \cdots & a_{-(n-1)} & a_0 & a_{-1} \\ a_{-1} & \cdots & \cdots & a_{-(n-2)} & a_{-(n-1)} & a_0 \end{pmatrix}. \quad (25.4)$$

Example 25.15. For $n = 4$, the eigenvalues of the matrix A in (25.4) are

$$\lambda_1 = a_0 - a_{-1} + a_{-2} - a_{-3}, \qquad \lambda_2 = a_0 + a_{-1} + a_{-2} + a_{-3},$$
$$\lambda_3 = a_0 - a_{-2} - i(a_{-3} - a_{-1}), \qquad \lambda_4 = a_0 - a_{-2} + i(a_{-3} - a_{-1}).$$

Theorem 25.11. For any two given circulant matrices A and B, the sum $A + B$ is circulant, the product AB is circulant, and $AB = BA$.

Example 25.16. For the matrices $A = \text{circ}(2, 1, 5)$, $B = \text{circ}(4, 3, -1)$, we have $A + B = \text{circ}(6, 4, 4)$ and $AB = BA = \text{circ}(22, 5, 21)$.

Problems

25.1. Use Theorem 25.1 to show that the following matrices are reducible:

$$A = \begin{pmatrix} 2 & 3 & 5 & 2 \\ 0 & 5 & 3 & 0 \\ 0 & 3 & 2 & 0 \\ 0 & 4 & 1 & 0 \end{pmatrix}, \quad B = \begin{pmatrix} 0 & 1 & 3 & 2 \\ 0 & 5 & 0 & 7 \\ 2 & 0 & 0 & 1 \\ 0 & 3 & 0 & 2 \end{pmatrix}.$$

25.2. Prove Theorem 25.2.

25.3. Use Theorem 25.2 to determine whether the following matrices are reducible or irreducible:

$$A = \begin{pmatrix} 0 & 2 \\ 3 & 4 \end{pmatrix}, \quad B = \begin{pmatrix} 3 & 2 \\ 0 & 4 \end{pmatrix}, \quad C = \begin{pmatrix} 4 & 0 & 7 \\ 11 & 8 & 0 \\ 0 & 5 & 8 \end{pmatrix}.$$

25.4. Prove Theorem 25.4.

25.5. Use Theorem 25.4 to show that the following matrices are invertible:

$$A = \begin{pmatrix} 7 & 2 & 3 \\ 1 & 5 & 2 \\ 3 & 4 & 8 \end{pmatrix}, \quad B = \begin{pmatrix} 5 & 4 & 1 \\ 3 & 3 & 0 \\ 0 & 6 & 6 \end{pmatrix}.$$

25.6. Prove Theorem 25.6.

25.7. Show that the matrices C and D in Example 25.1 are not monotone.

25.8. Let $A = (a_{ij})$ and $B = (b_{ij})$ be $n \times n$ monotone matrices. Show that if $A \geq B$, i.e., $a_{ij} \geq b_{ij}$, $1 \leq i, j \leq n$, then $A^{-1} \leq B^{-1}$.

25.9. Use Theorem 25.6 to show that the following matrices are monotone

$$A = \begin{pmatrix} 2 & 2 & -3 \\ -3 & 2 & 2 \\ 2 & -3 & 2 \end{pmatrix}, \quad B = \begin{pmatrix} 28 & -7 & 1 \\ -42 & -28 & 70 \\ 14 & 35 & -27 \end{pmatrix}.$$

25.10. Prove Theorem 25.9.

25.11. Find the eigenvalues and eigenvectors of the following matrices:

$$A = \begin{pmatrix} 1 & 3 & 0 & 0 \\ 3 & 1 & 3 & 0 \\ 0 & 3 & 1 & 3 \\ 0 & 0 & 3 & 1 \end{pmatrix}, \quad B = \begin{pmatrix} a_0 & 0 & a_{-2} & 0 \\ 0 & a_0 & 0 & a_{-2} \\ a_{-2} & 0 & a_0 & 0 \\ 0 & a_{-2} & 0 & a_0 \end{pmatrix}.$$

25.12. Show that $\text{circ}(1, -1, 2, 3)\text{circ}(4, 1, 5, -3) = \text{circ}(20, 6, 3, 6)$.

Answers or Hints

25.1.
$$
\begin{pmatrix} 2 & 3 & 0 & 0 \\ 3 & 5 & 0 & 0 \\ 1 & 4 & 0 & 0 \\ 5 & 3 & 2 & 2 \end{pmatrix}
=
\begin{pmatrix} 0 & 0 & 1 & 0 \\ 0 & 1 & 0 & 0 \\ 0 & 0 & 0 & 1 \\ 1 & 0 & 0 & 0 \end{pmatrix}
A
\begin{pmatrix} 0 & 0 & 0 & 1 \\ 0 & 1 & 0 & 0 \\ 1 & 0 & 0 & 0 \\ 0 & 0 & 1 & 0 \end{pmatrix}.
$$

$$
\begin{pmatrix} 5 & 7 & 0 & 0 \\ 3 & 2 & 0 & 0 \\ 1 & 2 & 0 & 3 \\ 0 & 1 & 2 & 0 \end{pmatrix}
=
\begin{pmatrix} 0 & 1 & 0 & 0 \\ 0 & 0 & 0 & 1 \\ 1 & 0 & 0 & 0 \\ 0 & 0 & 1 & 0 \end{pmatrix}
B
\begin{pmatrix} 0 & 0 & 1 & 0 \\ 1 & 0 & 0 & 0 \\ 0 & 0 & 0 & 1 \\ 0 & 1 & 0 & 0 \end{pmatrix}.
$$

25.2. Let A be an irreducible matrix and suppose that its directed graph G is not strongly connected. We suppose that G has n edges. Then, there are vertices v_i and v_j such that between them there does not exist any path. We denote with S the set of edges connected to v_j and with T the rest of the edges. It is clear that the sets S and T are non-empty, since $v_j \in S$ and $v_i \in T$. This implies that no edge $v \in S$ is connected with an edge $w \in T$, since otherwise $w \in S$, which is false. If we reorder the edges in the graph G and suppose that the first q edges are in S and the next $n-q$ vertices are in T, then we have $a_{rs} = 0$ for $r \in S, s \in T$. But this contradicts our assumption that A is irreducible. The converse requires a similar argument.

25.3. A irreducible, B reducible, C irreducible.

25.4. Assume that A is strictly dominated and noninvertible. Then, at least one of the eigenvalues of A, say, $\lambda_m = 0$. Let the eigenvector corresponding to λ_m be $u = (u_1, \cdots, u_n)$. Since $Au = \lambda_m u = 0$, it follows that $\sum_{j=1}^n a_{ij} u_j = 0$, $1 \le i \le n$. Let $\|u\|_\infty = \max_{1 \le i \le n} |u_i| = |u_k|$. Then, we have $a_{kk} u_k = -\sum_{j=1, j \ne k}^n a_{kj} u_j$, which gives $|a_{kk}| \le \sum_{j=1, j \ne k}^n |a_{kj}| |u_j / u_k|$, or $|a_{kk}| \le \sum_{j=1, j \ne k}^n |a_{kj}|$. But this contradicts our assumption that A is diagonally dominated.

25.5. Matrix A is strictly diagonally dominant and its inverse is $\frac{1}{187}\begin{pmatrix} 32 & -4 & -17 \\ -2 & 47 & -17 \\ -17 & -34 & 51 \end{pmatrix}$. Matrix B is diagonally dominant and irreducible and its inverse is $\frac{1}{12}\begin{pmatrix} 6 & -6 & -1 \\ -6 & 10 & 1 \\ 6 & -10 & 1 \end{pmatrix}$.

25.6. Let A be monotone and $A^{-1} = (b^1, \cdots, b^n)$. Then, $Ab^j = e^j \ge 0$, $1 \le j \le n$ implies $b^j \ge 0$, $1 \le j \le n$. Thus, $A^{-1} \ge 0$. Conversely, if $A^{-1} \ge 0$ and $Au \ge 0$, then $u = (A^{-1}A)u = A^{-1}(Au) \ge 0$.

25.7. In view of Theorem 25.6 it suffices to observe that not all elements of the matrices C^{-1} and D^{-1} are nonnegative:

$$C^{-1} = \begin{pmatrix} \frac{2}{3} & -\frac{7}{18} & -\frac{2}{27} \\ -1 & \frac{5}{6} & \frac{1}{9} \\ \frac{2}{3} & -\frac{5}{9} & \frac{1}{27} \end{pmatrix}, \quad D^{-1} = \begin{pmatrix} \frac{17}{18} & \frac{7}{9} & -\frac{11}{6} \\ \frac{1}{18} & \frac{2}{9} & -\frac{1}{6} \\ -\frac{7}{18} & -\frac{5}{9} & \frac{7}{6} \end{pmatrix}.$$

25.8. Follows from the identity $B^{-1} - A^{-1} = B^{-1}(A - B)A^{-1}$.

25.9. $A^{-1} = \frac{1}{5} \begin{pmatrix} 2 & 1 & 2 \\ 2 & 2 & 1 \\ 1 & 2 & 2 \end{pmatrix}$, $B^{-1} = \frac{1}{308} \begin{pmatrix} 11 & 1 & 3 \\ 1 & 5 & 13 \\ 7 & 7 & 7 \end{pmatrix}$.

25.10. From Theorem 17.5, it suffices to note that $A^{-1} = \sum_{k=1}^{\infty} B^k$ and $B \geq 0$.

25.11. For A in Example 25.14 take $b_0 = 1, b_1 = 3$.
For the matrix B, $\lambda_1 = \lambda_2 = a_0 + a_{-2}$, $\lambda_3 = \lambda_4 = a_0 - a_{-2}$, $v^1 = (0,1,0,1)^t$, $v^2 = (1,0,1,0)^t$, $v^3 = (0,-1,0,1)^t$, $v^4 = (-1,0,1,0)^t$.

25.12. Verify by direct multiplication.

Bibliography

[1]. R.P. Agarwal, *Difference Equations and Inequalities: Second Edition, Revised and Expanded*, Marcel Dekker, New York, 2000.

[2]. R.P. Agarwal and D. O'Regan, *An Introduction to Ordinary Differential Equations*, Springer–Verlag, New York, 2008.

[3]. A. Albert, *Regression and the Moore–Penrose Pseudoinverse*, Academic Press, New York, 1972.

[4]. A. Ben–Israel and T.N.E. Greville, *Generalized Inverses: Theory and Applications*, Wiley, New York, 1974.

[5]. G. Birkhoff and T.C. Bartee, *Modern Applied Algebra*, McGraw–Hill, New York, 1970.

[6]. G. Birkhoff and S. Mac Lane, *A Survey of Modern Algebra*, Macmillian, New York, 1977.

[7]. C.T. Chen, *Introduction to Linear System Theory*, Holt, Rinehart and Winston, Inc., New York, 1970.

[8]. F.R. Gantmacher, *Applications of the Theory of Matrices*, Interscience, New York, 1959.

[9]. P.R. Halmos, *Finite–Dimensional Vector Spaces*, Van Nostrand, New York, 1958.

[10]. K. Hoffman and R. Kunze, *Linear Algebra*, 2nd ed., Prentice–Hall, Englewood, N.J., 1971.

[11]. A.S. Householder, *The Theory of Matrices in Numerical Analysis*, Blaisdell, New York, 1964.

[12]. P. Lancaster, *Theory of Matrices*, Academic Press, New York, 1969.

[13]. C.L. Lawson and R.J. Hanson, *Solving Least Squares Problems*, Prentice–Hall, Englewood, N.J., 1974.

[14]. S.J. Leon, *Linear Algebra with Applications*, Macmillian, New York, 1980.

[15]. M.Z. Nashed, *Generalized Inverses and Applications*, Academic Press, New York, 1976.

[16]. B. Noble, *Applied Linear Algebra*, 2nd edition, Prentice–Hall, Englewood, N.J., 1977.

[17]. B.N. Parlett, *The Symmetric Eigenvalue Problem*, Prentice–Hall, Englewood, N.J., 1980.

225

[18]. G. Strang, *Linear Algebra and its Applications*, 2nd edition, Academic Press, New York, 1976.

[19]. R.M. Thrall and L. Tornheim, *Vector Spaces and Matrices*, Wiley, New York, 1957.

[20]. R.A. Usmani, *Applied Linear Algebra*, Taylor & Francis, New York, 1986.

Index